AF719282

SURFACE MINE RECLAMATION MANUAL

WITHDRAWN

SURFACE MINE RECLAMATION MANUAL

E.S. Lyle, Jr.
School of Forestry
Auburn University

Elsevier
New York • Amsterdam • London

S
621.5
.S8
L95
1987

Elsevier Science Publishing Co., Inc.
52 Vanderbilt Avenue, New York, New York 10017

Sole distributors outside the United States and Canada:

Elsevier Science Publishers B.V.
P.O. Box 211, 1000 AE Amsterdam, The Netherlands

©1987 by Elsevier Science Publishing Co., Inc.

This book has been registered with the Copyright Clearance Center, Inc. For further information please contact the Copyright Clearance Center, Salem, Massachusetts.

Library of Congress Cataloging in Publication Data

Lyle, E. S.
Surface mine reclamation manual.

Includes index.
1. Reclamation of land. 2. Strip mining—Environmental aspects.
3. Revegetation. 4. Soil conservation.
I. Title.
S621.5.S8L95 1986 631.6'4 85-31153
ISBN 0-444-01014-9

Current printing (last digit):
10 9 8 7 6 5 4 3 2 1

Manufactured in the United States of America

This book is dedicated to the owners and personnel of Drummond Coal Co., whose progressiveness, cooperation, and encouragement made this book possible.

CONTENTS

PREFACE

The purpose of this manual is to provide a single source that discusses most of the factors necessary for revegetation and control of erosion and sediment on surface mines. The manual is divided into three parts. The first part discusses revegetation theory. The second part discusses revegetation practice, and the third part discusses soil erosion and sediment control. The material in this manual has been gleaned from many sources. However, there has been an effort to condense, simplify, and present this material in a manner that would be immediately useful to a surface mine reclamationist.

It is recognized that many persons doing reclamation, or interested in reclamation, do not have formal training in all phases of plant and soil sciences. With this fact in mind, an attempt has been made to avoid terms that require detailed explanations. Also, an attempt has been made to discuss only the factors that reclamationists would manipulate or need to understand in the course of their duties. Naturally there are many factors other than the ones discussed in this manual that influence revegetation and erosion control. References and suggested readings are given for those who would like to delve further into the factors of reclamation.

The author is indebted to many people and organizations for the information in this book. He is particularly indebted to Mr. Dwight R. Hicks, Mr. Bill Henslee, Dr. David S. Ralston, and the United States Soil Conservation Service.

INTRODUCTION

Surface mine reclamation is often thought of in terms of revegetation, and this is not unreasonable. Planned revegetation of a minesoil controls erosion, puts the soil back into useful production, and creates an aesthetically pleasing terrain. The short-term goal of revegetation is to control soil erosion quickly with fast-growing plants. The long-term goal should be to establish a crop that benefits the landowner and the public. These long-term crops could be forage crops, fruit orchards, forest trees, or row crops. Of course, there are other long-term uses for reclaimed mines such as family housing, industrial parks, and recreational areas.

Soil is the medium necessary for revegetation, and the soil must be amenable to plant growth. Soils are formed over long periods of time from geologic materials such as stone or clay. These geologic materials are often poor for plant growth. They must often evolve through weathering, biologic processes, and leaching processes before they become productive. The mine overburden can be considered geologic material from which soil is formed. Therefore surface mines can be considered new soils that will go through the complete aging process, even though the new material may be suitable for plant growth before aging.

It seems reasonable to give the soil material that is to be vegetated a special name, as many of its characteristics are derived from human rather than nature's activities. For the purposes of this book the material deposited by the dragline or shovel will be called overburden. The soil material removed before the mining operation and replaced on the graded overburden will be called topsoil, and the combination of topsoil over overburden will be called minesoil.

The new overburden goes through a weathering process, and organic matter increases to form a young soil. Eventually the soil characteristics controlled by the overburden character become lessened due to weathering, and the soil now has new characteristics formed by the local environment. This soil is usually in equilibrium with the local environment, and the soil characteristics change slowly after an equilibrium is established. This soil is now considered a mature soil. Under these circumstances the reclamationist is often placing an old and infertile topsoil over a material that may eventually become highly productive soil. The eventual effect of this type of activity is not known, but it seems likely that there are circumstances in which topsoil should be replaced, ones in which topsoil should not be replaced, and even ones in which the topsoil and overburden should be mixed.

It also seems likely that on many occasions reclamationists can improve the topography of the land, fertility of the soil, and economics of the locality. However, it is recognized that this is often difficult to accomplish under the restrictions set by government regulations, company policies, lack of funds, and time.

The field of reclamation is advancing at a fast rate. The methods used in reclamation are not new, but most of these methods have required varying degrees of modification to meet the unusual aspects of surface mine reclamation. These modifications are continuing and gradually leading to more successful reclamation. The methods presented in this manual are not intended to be accepted as indisputable facts. Reclamationists must place this information and any other information under careful scrutiny and then, in the light of experience, use it in the manner best suiting their objectives and conditions. It is hoped that the theory and practice presented here will be comprehensive and used by reclamationists to develop new methods and modify present methods to meet individual needs.

SURFACE MINE RECLAMATION MANUAL

PART I

ENVIRONMENTAL FACTORS AND THEORIES OF PLANT CULTIVATION

CHAPTER 1

PRIMARY FACTORS AFFECTING SEED GERMINATION, PLANT ESTABLISHMENT, AND GROWTH

The plants grown for reclamation purposes are 94 to 99.5 percent carbon, hydrogen, and oxygen, and these are obtained by the plants from air and water. Carbon and oxygen are obtained from the atmosphere and hydrogen from water in the soil. The other 6 to 0.5 percent of the plant is composed of elements (nutrients) obtained from the soil. This small percent of the plant obtained from soil is often the factor that limits plant growth, but this factor usually can be controlled through liming or mineral fertilization.

There are approximately eight environmental factors that have a major effect on plants. The reclamationist can have significant control over four of these factors but has much less control over the other four. They are all factors with which a practicing reclamationist is familiar. The purpose of this chapter is to organize these factors and look at the interrelationships among them. These interrelationships are probably extremely complex, but this does not matter greatly for reclamation purposes. Once you get one of the factors into the condition necessary for good plant growth, the interrelated factors also tend to become modified to produce good plant growth.

The eight factors that have a primary effect on all plant growth are:

Factor	Where They Function
1. Sunlight	Atmosphere and soil
2. Temperature	Atmosphere and soil
3. Carbon dioxide content	Atmosphere and soil
4. Animals and other plants	Above soil and in soil
5. Water	Primarily in soil
6. Oxygen	Soil
7. Nutrients	Soil
8. Acidity	Soil

SUNLIGHT

Sunlight is absolutely necessary for the growth of all plants used in reclamation, but there is very little that the reclamationist can do to control sunlight. The one exception is in the germination of seeds. A few plants have seeds that require light to germinate, whereas most seeds do not. Therefore there may be a need to expose seeds to sunlight. But, if possible, it is usually better to cover the seeds with mulch or soil. Many years of experience and research have provided the agriculturist with the necessary information regarding seed cover and the depth to which specific seeds should be planted.

The other situation in which the effects of sunlight may be controlled is the number of plants grown on a specific area of soil. If too many plants are crowded on an area, they may shade each other and prevent the shaded plants from manufacturing maximum amounts of food, thereby reducing the crop yield. Also, if a mixture of plant species is grown on an area, the taller plants may shade the shorter plants and affect some aspect of the shorter plants' growth. Although shading through overcrowding can reduce a plant's growth, it is even more likely that the loss of growth will be due to excessive root competition for moisture and nutrients. It has been well proved that such excessive competition among plants for moisture and nutrients will cause a loss of growth in individual plants.

TEMPERATURE

Temperature is important for seed germination and plant growth, but again the reclamationist has little control over atmosphere or soil temperatures. Soil temperature can be controlled to a degree by using mulches to block the sun's rays. Temperature has a direct effect on many plant functions such as respiration, photosynthesis, transpiration, and the absorption of water and nutrients by roots. Respiration is the production of energy by the plant for the purpose of carrying out life processes. Photosynthesis is the process by which plants use sunlight to manufacture the food for respiration and plant growth. Transpiration is the process of releasing water vapor from plant leaves into the atmosphere.

As the temperature of soil water approaches freezing, the water becomes more viscous and therefore more difficult for the plant root to absorb the water. As the temperature of water rises from 0°C to about 65°C, it becomes increasingly easier for the plant root to absorb water. Above 65°C there is little difference in the absorption rate. As the temperature of soil water near the soil surface rises there is a tendency for the water to evaporate more rapidly and pass from the soil into the atmosphere. Again mulching modifies conditions and helps to prevent this rise in temperature and subsequent loss of soil water.

Low soil temperatures also retard the uptake of nutrients by plant roots. This is due to the slowing of water uptake and life process activities of the plant at lower temperatures.

CARBON DIOXIDE

Only about 0.23 percent of the atmosphere is composed of carbon dioxide. However, this small percent is vital for plant growth. Carbon dioxide is used by the plant through photosynthesis to form the organic molecules that result in plant growth. The carbon dioxide in these organic molecules is returned to the atmosphere when the plant dies and decays. It is known that increasing carbon dioxide in the atmosphere will increase plant growth. This is not true in the soil. Increased carbon dioxide content of the soil means reduced oxygen content, and oxygen is needed by plant roots to absorb water and nutrients.

ANIMALS AND OTHER PLANTS

Animals and other plants can have a disastrous effect on growth of the desired plants. Many of the disease organisms that injure plants are lower forms of plant life such as fungi. Insects cause many plant problems by feeding on plant parts, and higher animals such as rabbits and deer destroy plants through grazing. The effects of other organisms on plants are not always detrimental. Some bacteria are used to inoculate the roots of legumes in order to convert atmospheric nitrogen into a form of nitrogen that plants can use in their growth processes. Also certain fungi combine with the roots of some higher plants in order to benefit both the fungi and the higher plants.

Sunlight, temperature, carbon dioxide, and organisms are all extremely important for plant growth. However, these items are either not easy to control or have not thus far caused a serious problem in reclamation. Some of the growth factors such as nitrogen-fixing bacteria, fungi-root associations, and soil temperature will be discussed again under other headings.

The last four growth factors—soil water, soil oxygen, soil nutrients, and soil acidity (pH)—will be discussed in more detail because they are factors that the reclamationist can control to a significant degree.

SOIL WATER

Soil water is the factor with which the agriculturist has been most concerned throughout history and even before. The growth of all plants used in reclamation is more or less proportional to the amount of water in the soil. There is little growth on dry soils. Growth increases as the dry soil becomes wetter until the soil water reaches a level at which oxygen is driven from the soil. At this point most plants stop growth, just as they do when the soil is too dry. Many soil factors must be in balance with each other to obtain satisfactory plant growth. Fortunately these balances do not have narrow limits, and acceptable growth can be obtained over a reasonably wide range of factor variation. Water is used by the plant to build tissue and to transport raw materials to the leaves for food manufacture as well as to transport the manufactured

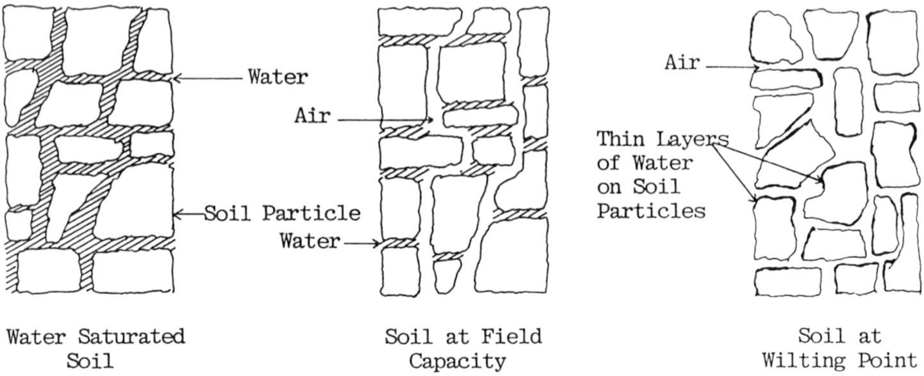

Figure 1. Relative amounts of water and air in a soil at three soil moisture conditions.

food to other parts of the plant. A lack of water can cause plant cells to fail to multiply or increase in size. Of course, this means a reduction or cessation of plant growth.

Soil water also has an effect on the uptake of nutrients from the soil. As the soil moisture is increased from the lowest point at which plants can live to the ideal soil moisture content, there is a corresponding increase in nutrient uptake. However, when you increase soil water above the ideal content, the amount of soil oxygen is decreased and the soil's ability to take up nutrients and water is decreased.

Finally, soil moisture is needed at the soil surface for seed germination and the establishment of the resulting seedling. This is a point at which sunlight and temperature also play an important part. Too much heat from sunlight can dry the soil surface and prevent germination or kill the newly germinated and tender seedling. Mulches help to control this problem.

All soil moisture can be classified as being in one of three states. Immediately after a heavy rain, the soil pores will be filled with water and practically all air will be excluded. The water will be moving downward into the lower soil layers. At this point the soil is saturated with water (Figure 1). After a period of time the downward movement of water will cease, and air will move back into the larger soil pores. This soil moisture state is called field capacity (Figure 1). This is probably the point at which plants can most easily obtain water. The soil will continue to dry due to plant use and drying at the soil surface. Eventually the soil will dry to the point at which plants can no longer absorb the remaining moisture. This point is called the wilting point (Figure 1). A short period at the wilting point will kill young plants, and a longer period will kill mature plants.

There are four soil characteristics that have a major effect on the ability of soil to absorb water, store water, and allow plant roots to absorb water. These four characteristics are: (1) soil texture; (2) soil

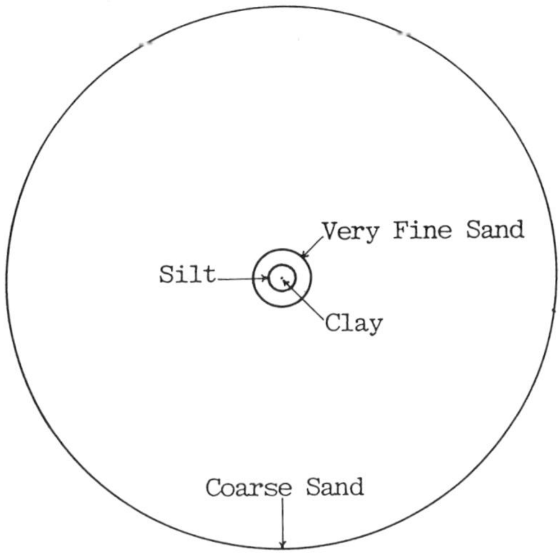

Figure 2. Relative soil particle sizes.

structure; (3) soil organic matter; and (4) soil depth. There are many interactions among these four characteristics that have important effects on soil water and plant growth. Fortunately, improvement in any one of the characteristics usually improves one or more of the other characteristics. For instance, decaying organic matter will improve the structure of a heavy clay by forming soil particles into small soil clods with spaces between the clods. These spaces allow water and air to move into the soil. Without the organic matter, the heavy clay might form a solid layer at the soil surface that would be impenetrable to water and air.

Soil Texture

Soil texture is an easily determined soil characteristic. It is determined by the relative percents of sand, silt, and clay found in a soil. The only thing that determines whether a soil particle is sand, silt, or clay is the size of that particle. Sand is any soil particle between the sizes of 2.0 mm and 0.05 mm. Silt is between 0.05 mm and 0.002 mm in size, and clay is anything less than 0.002 mm. The size of a soil particle also determines whether or not material can be called soil. Any material greater than 2.0 mm in diameter is not considered soil. There is no lower size limit for soil. Figure 2 displays the relative soil particle sizes. If a soil was composed of 40 percent sand, 40 percent silt, and 20 percent clay, the texture classification would be loam. If the percentages were 60 percent sand, 30 percent silt, and 10 percent clay, the texture would be sandy loam. Soil with 60 percent or more clay is classified as clay regardless of the particle sizes of the rest of the soil.

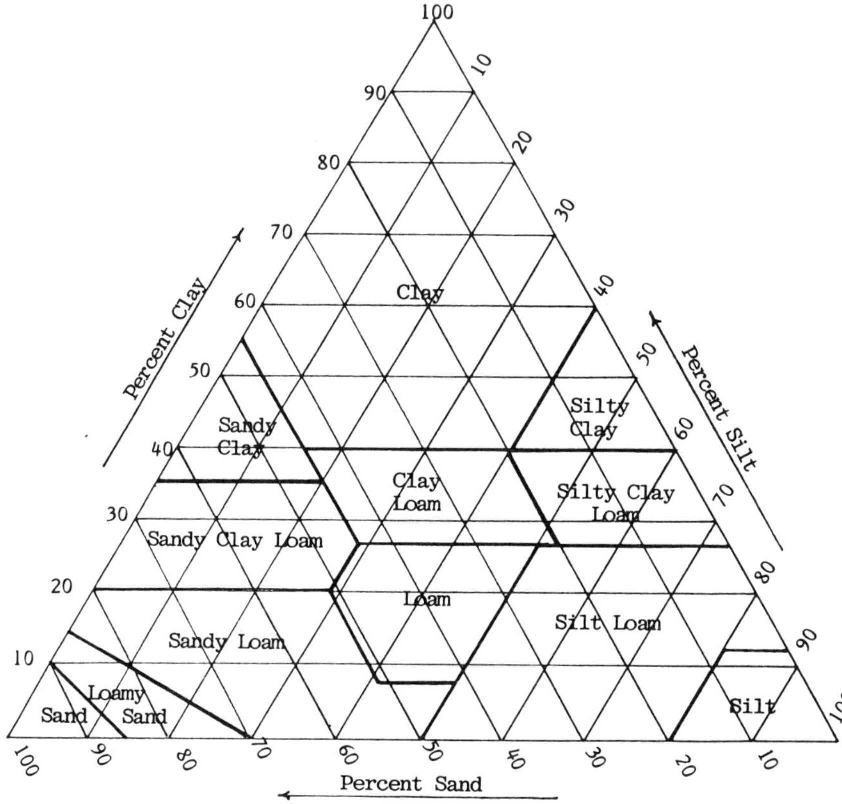

Figure 3. Soil texture nomograph.

Figure 3 is the texture triangle nomograph used to determine soil texture when the amounts of sand, silt, and clay in the soil are known. The amounts of sand, silt, and clay can be determined with simple laboratory procedures, which are done when positive identification of texture is needed. However, approximate texture can be determined in the field by simply wetting the soil and rubbing it between the thumb and fingers. This technique can be learned quickly and is usually all that is needed for a useful texture determination.

Soils that have a high sand content are said to have a coarse texture, and those with high silt or clay contents are said to have a fine texture. Coarse soils such as sands allow water to enter the soil freely, but they may also allow water to pass through the soil without retaining enough for plant use. On the other hand, fine soil particles such as clays have a tendency to prevent the movement of water from the soil surface down into the soil. Once water enters a clay soil, the clay retains the water well. In many cases clay holds much of the soil water so tightly that it is unavailable to roots. Intermediate textures, such as

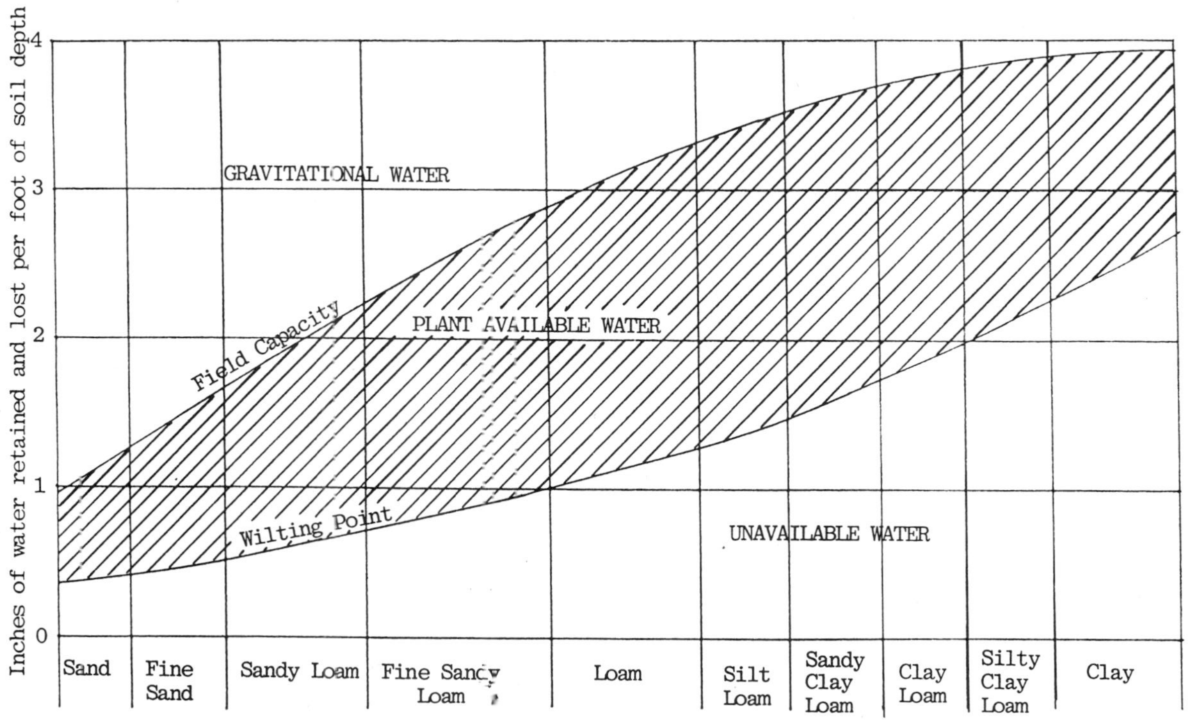

Figure 4. Effect of texture on a soil's ability to hold water. *Source:* U.S.D.A., Forest Service, *Handbook on Soils,* 1961.

silt loams, sandy loams, sandy clay loams, and loams, have a more desirable combination of characteristics than sand or clay and allow water to enter the soil. However, they also retain available water in the root zone for plant use. Figure 4 shows the approximate effect of texture on a soil's ability to hold water. The sandy soils do not hold water well and much of the water passes downward through the soil. Water that moves downward and is not held by the soil is called gravitational water. Sandy soils do not hold much available or unavailable water. Silt loams and silts hold the most available water for plant growth even though much of the total water held is unavailable. As the clay content of soil increases, the soils hold more and more water. However, most of the water is held so tightly by the clay that the plants cannot absorb the water. The preceding statements apply to the effect of texture alone on soil moisture relationships. Soil structure can modify the effect of soil texture and significantly change soil moisture relationships. Therefore, soil structure can cause a wide range in available and unavailable water content for any given soil texture.

This ability of a soil to allow water to enter and then store the water comes about because of the number and sizes of pores in the soil. These open spaces or pores are just as important as the solid portion of the soil. Without pores, there would be no water or oxygen in the soil and both of these are vital for plant growth. Two general types of pores exist in most soils. These two types are macropores and micropores. However, there is no sharp line of distinction between macropores and micropores. A macropore is a pore large enough to allow fast movement of air and water through the soil. In micropores air movement is much slower and water movement is a slow capillary movement. The need for both water and air reveals the old agriculture problem of trying to maintain a useful balance among growth factors in order to obtain good plant growth. Macropores are needed to get water and air into the soil and well distributed throughout the root zone. However, micropores are also needed in order to hold some of the water by capillarity for the roots to utilize until the next rain.

Sandy soil usually has less total pore space than clay soil, and most of the pores are macropores. This means that the movement of air and water through the soil is rapid, but little water is held in the root zone. Fine-textured soil such as clay usually has a greater amount of pore space than sandy soil, and most of the pores tend to be micropores, so the water and air movements are slow. Under these circumstances, the clay soil will retain water for the plant roots, but it may retain too much water. The spore spaces then may completely fill with water, and there will not be enough oxygen in the root zone.

A good balance of solid and pore space in a soil is about 50:50. Fifty percent of the soil should be solid material and 50 percent pore space, with the pore space equally divided between macropores and micropores (Figure 5). Under these conditions the macropores tend to fill with air shortly after a rain, and the micropores fill with water. This situation provides water and oxygen for optimum root performance and plant growth.

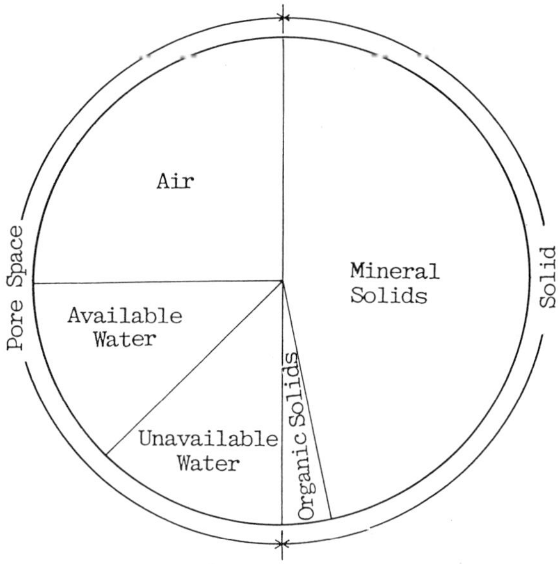

Figure 5. Proportions of solids, water, and air that would provide a good plant growth medium.

Soil Structure

The second soil characteristic affecting soil water content is soil structure. Structure modifies soil properties and almost always does so in a beneficial way. Soil structure is the aggregation of soil particles into clusters of particles that produce a characteristic form. Figure 6 illustrates soils with and without structure. These are examples of typical structures, and there are many gradations among them. Initially most soils on reclaimed surface mines will have no structure. Any structure in the original soils will have been destroyed by moving the soil from one place to another and by compaction with equipment in the replacement and grading processes.

Unstructured, sandy soil is classified as single-grained. Clay soils without structure tend to be large expanses of soil material with few macropores. These soils are classified as massive. Both single-grained and massive soils occur in minesoils.

There are several processes or factors that will restore structure to a minesoil. Wetting and drying moves the soil particles back and forth and causes some particles to move close together and other particles to move further apart. This sort of action causes aggregates to form with spaces between the aggregates. Freezing and thawing and the movement of roots through the soil have much the same effect as wetting and drying.

Organic matter in the soil may have more to do with the development of structure in a soil than any other factor. The mulch used on a minesoil and the stems, leaves, and roots of plants growing on a minesoil

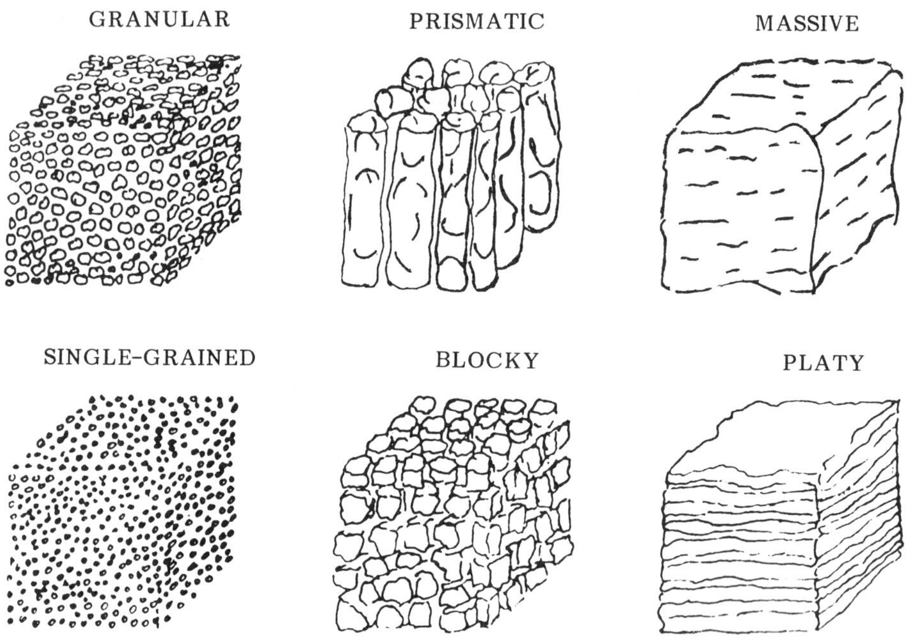

Name	Unit Size	Where Found	Permeability	Comments
Granular	1/2" or less diameter	Surface soil	Rapid	High in organic matter
Prismatic	Up to 6" in diameter	Subsoil	Moderate	
Massive	Unlimited	Surface soil or subsoil	Slow	Unstructured unweathered clay
Single-grain	Sand size	Surface soil or subsoil	Rapid	Unstructured sands
Blocky	Up to 4" in diameter	Subsoil	Moderate	
Platy		Surface soil	Slow	Virgin surface soils

Figure 6. Structured and unstructured soils. *Source:* U.S.D.I. and U.S.D.A., *Irrigation on Western Farms.* Agr. Inf. Bul. No. 199, 1959.

all contribute to the organic matter in a soil. Microorganisms that decay organic matter and the products produced by decay cause sands to aggregate and form granules. These products also aid in breaking massive clays and forming them into granules.

Some of the nutrients in a soil can influence the formation of granules. If a soil has an excess of sodium, the soil granules tend to fall apart, with sands becoming single grains and clays becoming massive. However, calcium tends to make clays or humus aggregate and form granules.

Finally, plowing and other tillage operations can have a short-term beneficial effect when performed under the proper soil moisture conditions. However, if the soil is too dry, much of the soil structure will be destroyed by crushing the granules. If the soil is tilled while too wet, large clods will be formed or the soil will be unnecessarily compacted. The overall effect of tillage on soil moisture relations is often detrimental. It hastens the loss of organic matter, disrupts original soil structure, and compacts the soil when equipment wheels or tracks pass over the soil. Of course, some tillage is usually necessary, but it should be done under the proper moisture conditions to prepare a favorable seedbed and to do as little damage as possible to the existing structure.

Soil Organic Matter

Organic matter has already been mentioned in connection with other soil factors, but it has many effects on soil water and should be discussed as an important factor. Organic matter is the rotting or decomposing remains of plants or animals. The stage of decomposition can be anything from fresh material that has just fallen to the ground to material that has broken down into liquids, slimes, and gums. Organic matter not only has much to do with soil water, but it is also important in the soil nutrient cycle.

Some of the more important effects of organic matter on soil water are as follows:

1. Fresh organic matter on the soil surface reduces the impact of falling raindrops. If the impact of raindrops is not reduced, they will strike the soil surface with enough force to move soil particles into and seal soil pores and cracks that would have allowed water to infiltrate the soil.
2. Fresh material also slows the lateral movement of water over the soil surface. If water is allowed to move at a fast rate over a soil surface, it will carry soil particles downhill. Therefore, any reduction in the speed of flowing water reduces the amount of soil erosion.
3. Fresh material at the soil surface reduces the loss of water from the soil surface because it shades the surface, reduces soil temperatures, and blocks the movement of drying winds over the soil surface.

4. Fresh material on the soil surface provides conditions that allow the soil surface to remain moist. This is beneficial for germinating seeds that are sown at the soil surface.
5. Fresh organic material supplies food for animals such as insects, worms, and rodents. These animals burrow into the soil, and their burrows help supply the soil with water and oxygen.
6. When the fresh material is decomposed it releases carbon dioxide, which forms carbonic acid. This helps to decompose minerals and release nutrients for plant growth.
7. Decomposed organic matter produces slimes and gums. These slimes and gums are invaluable in the formation and stabilization of soil structures that form cracks for rain and air to move into the soil.
8. Decomposed organic matter itself will absorb and hold water and nutrients for plant use. This is particularly important in sandy soils where water and nutrients would otherwise move through the soil and out of the root zone.

In the eastern U.S. it is difficult, if not impossible, to maintain organic matter in a soil if the soil is farmed continuously with row crops. The continuous tillage in the cultivation of row crops causes an accelerated decomposition of organic matter as it is added to the soil. However, if grasses and legumes are grown as pasture or forage crops, the organic matter can usually be maintained at a reasonable level.

Soil Depth

The final soil characteristic affecting the availability of soil water to plants is soil depth. Soil depth can be defined as the distance from the soil surface down to anything that prevents a plant root from growing and absorbing water and nutrients. A layer of rock, toxic chemicals, tight clay, or a water table can be the factor that prevents roots from penetrating deeper into a soil. It is obvious that a shallow soil (eg, 10 inches of soil over a layer of sandstone) cannot absorb or store as much water as 3 feet of the same soil over a layer of sandstone. Under present reclamation regulations, it is likely that some layers of minesoils will be compacted. This compaction may cause difficulties for rain penetration, water storage, and plant root penetration. However, a minesoil is unlikely to have any of the other soil depth limiting factors except toxic elements.

Most plant roots will not penetrate a soil layer saturated with water such as a water table. This is true because such soil would not be likely to have enough oxygen to allow roots to absorb water and nutrients. Any factor or group of factors that limit the depth to which roots can penetrate limits the amount of water and nutrients available for plant use.

Soils can be too deep. This is particularly true when soils are sandy or coarse-textured and there is no soil horizon that has properties to stop water from moving quickly below the zone where roots can absorb the needed water. Such soils would support more plant growth if a finer textured, stony, or compacted soil horizon were present 3 to 6 feet below the soil surface to keep soil water from moving too deeply and too quickly.

SOIL AIR

The amount of water held by the soil for plant use is extremely important, but so is the amount of air. The amount of air in a soil is determined to a large degree by the amount of water in the soil. As water moves into a soil it will displace the air. When rainwater moves down through a soil it first moves out of the large soil pores (macropores) and tends to remain in the small or capillary pores (micropores). This means that as the soil dries the air will move into the large pores first and then into increasingly smaller pores.

There are two soil conditions that usually result in poor soil aeration. The first occurs in compacted soil where there are few or no large soil pores. A massive soil would be an example of this condition. The second is any situation in which the soil pores are filled with water. This can occur immediately after a rain or when the soil is flooded for some other reason such as a high water table. Well-drained soil with good structure is likely to be well-aerated and will not restrict plant growth. However, a soil with a compacted layer near the surface or a high water table will prevent proper soil aeration and thereby restrict root development.

NUTRIENTS

Plants must have about 20 different nutrient elements to grow and develop properly. Not all plants use all of these elements, but each of these elements has been found essential for at least one plant. At most we are only concerned with 15 of these elements, and about 95 percent of the time we are only concerned with six of the elements. These six elements are nitrogen, phosphorous, potassium, sulfur, calcium, and magnesium. Table 1 lists the 15 elements that are most likely to be needed, along with their chemical symbols.

Most of a plant is composed of carbon, hydrogen, and oxygen, but these elements come from air and water so the reclamationist has little control over the supply. Also, there are several elements that plants use in such small amounts that they are usually present in adequate amounts and do not have to be added. The elements that do need to be added in many soils are nitrogen, phosphorous, potassium, calcium, magnesium, and sulfur. Most reclamationists add a fertilizer containing nitrogen, phosphorous, and potassium. The compound often used to add phosphorous to the fertilizer contains sulfur and if dolomitic

Table 1. Source of 15 Elements Used by Plants

Carbon dioxide from atmosphere	Soil water	Oxygen in atmosphere	Soil water & soil solids
Carbon (C)[1]	Hydrogen (H)	Oxygen (O)	Nitrogen (N) Phosphorous (P) Potassium (K) Calcium (Ca) Magnesium (Mg) Sulfur (S) Iron (Fe) Zinc (Zn) Copper (Cu) Manganese (Mn) Molybdenum (Mo) Boron (B)

[1]Chemical abbreviations for elements given in parentheses.

limestone is used for liming, calcium and magnesium are added to the soil. Therefore, it is convenient to add six of the most needed elements in the normal process of preparing a soil for revegetation.

It is useful to think of the movement of plant nutrients as the cycle shown in Figure 7. The cycle begins with the soil, and it should be remembered that soils have inherent nutrients. Some of these nutrients are available to plants, and some are not available. There is usually a constant movement of nutrients from available to unavailable forms, depending on soil conditions. Nutrients can be removed from the soil by soil erosion, microorganisms, or plant roots, and they can be leached below the rooting zone by soil water. Also nitrogen is sometimes lost when it converts to a gaseous form and escapes to the atmosphere. For all practical purposes, nutrients can be removed from the soil by changing from an available chemical form that roots can absorb to an unavailable form that roots cannot absorb. Nutrients are added to the soil when these unavailable forms change into an available form due to some activity such as a change in soil pH or the action of microorganisms. Nutrients are added directly to the soil when chemical fertilizers are applicd, when organic matter decomposes, when rain absorbs chemical compounds from the atmosphere, and when soil microorganisms die and release their stored nutrient elements back to the soil.

Plants growing on a minesoil utilize the water and nutrients of the minesoil to carry on their life processes. If the plants are harvested by humans or eaten by animals and removed from the minesoil, a significant amount of nutrient material is removed from the cycle. If the plants are allowed to die and remain on the minesoil, they return their stored nutrients to the minesoil. Of course, any organic mulches

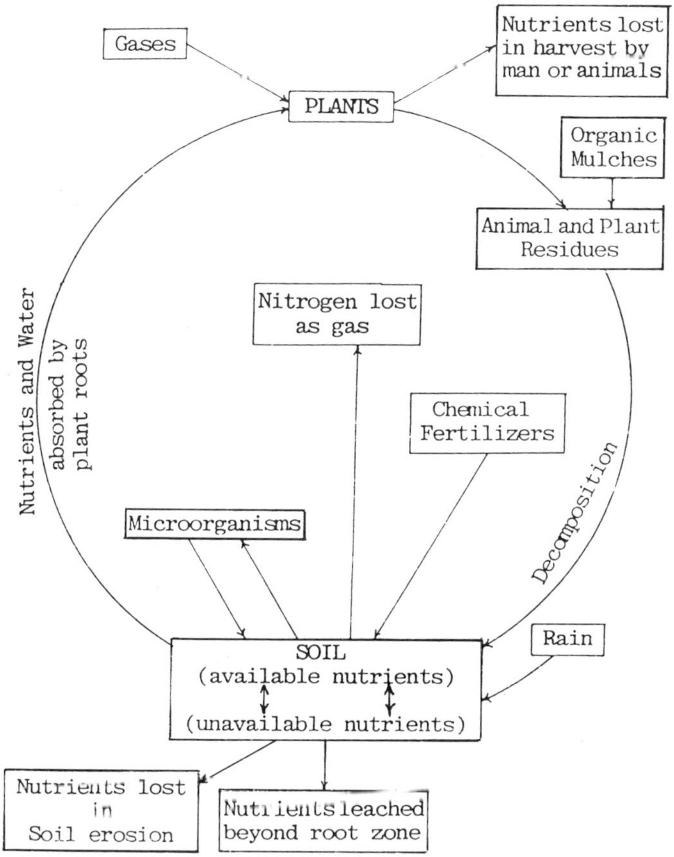

Figure 7. Nutrient cycle among plants, soil, and atmosphere.

such as straw also add some nutrients to the minesoil when they decompose.

There is one other aspect of soil nutrients and plant growth that the reclamationist needs to understand. If there is a deficiency of one nutrient needed for plant growth, the addition of some other nutrient will not increase plant growth significantly. The deficient nutrient must be added to the soil in order to increase plant growth. Once the deficient nutrient has been added there will be an increase in growth until some other nutrient becomes deficient. If a deficiency of two nutrients is limiting growth, then adding one of the nutrients may increase growth slightly. However, if both nutrients are added there will be a far greater increase in growth.

The above situation is true because the lack of any single essential nutrient will keep a plant from growing. Once this essential nutrient is added, the plant will increase its rate of growth until the next nutrient

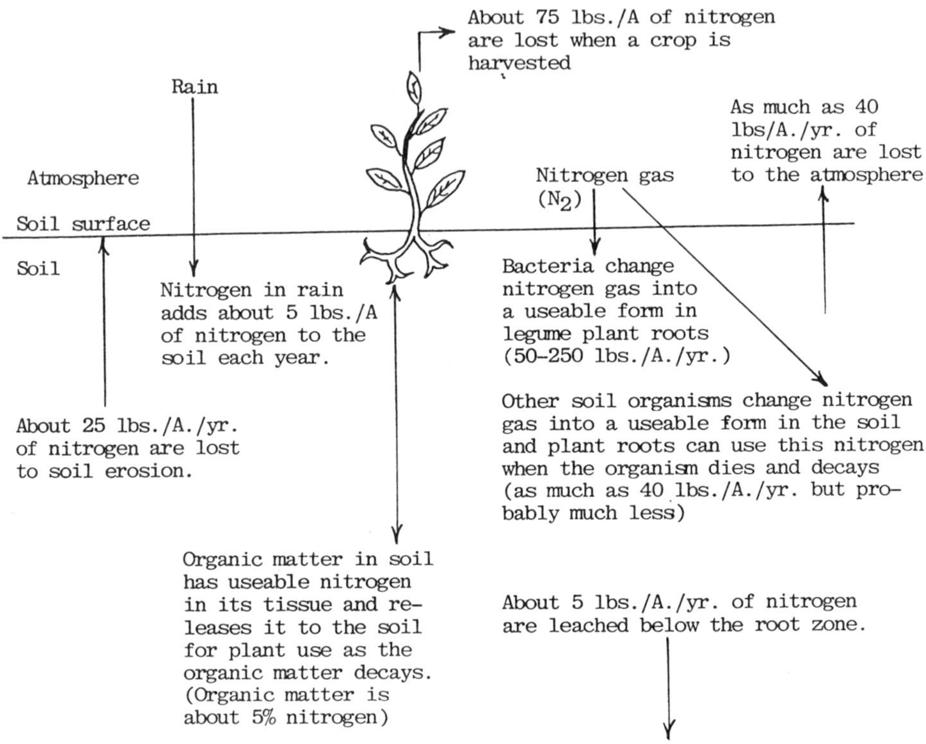

Figure 8. Possible nitrogen activities in a soil containing organic matter.

that is in short supply causes the rate of growth to slow. If this second nutrient supply is increased, the rate of plant growth will increase again until a third nutrient becomes limiting. Of course, there is a limit to the amount of plant growth that can be obtained by increasing the nutrient supply and excessive increases may even be toxic to the plant.

Nitrogen

Nitrogen is one of the most abundant elements in our world. Approximately 78 percent of the air is composed of nitrogen. This is about 70 million pounds of nitrogen for every acre of the earth's surface, including oceans. Unfortunately, this nitrogen is not in a form that can be directly utilized by most plants. Also, nitrogen is not stored in the soil except in organic matter. Organic matter is approximately 5 percent nitrogen and it is unfortunate that most new minesoils contain very little organic matter. However, minesoils do acquire organic matter and the resulting nitrogen from decomposing vegetative covers.

Plants that are slightly deficient in nitrogen do not have a normal amount of growth. As the deficiency becomes greater, the plants become noticeably stunted and yellowish in color. The yellow color appears on

Table 2. Possible Nitrogen Gains and Losses in a Soil Containing Organic Matter[1]

Gains (lb/A/yr)		Losses (lb/A/yr)	
Legume fixation	50[2]–250[3]	To atmosphere	40
Other soil organism fixation	40	Leaching	5
Rain	5	Harvesting	75
		Erosion	25
Total	95–295	Total	145

[1]The nitrogen in organic matter tends to cycle between living and dead organic matter. However, the nitrogen in the organic matter cycle can be lost or gained through the agents shown above.

[2]Forest.

[3]Field crop.

the older or lower leaves before appearing on the newer or upper leaves. If the plant is a grass, the tip of the leaf turns brown and the browning progresses along the midrib until the middle and lower parts of the leaf turn brown. The addition of nitrogen to a soil in which a nitrogen-deficient young plant is growing is quickly noticed. The plant takes on a deeper green color and vegetative growth is noticeably accelerated.

The nitrogen used by plants on minesoils usually comes from fertilizers, legumes, and the organic matter supplied by crops. If the crops are not harvested, much of the nutrient material that the plants have taken from the soil will be returned to the soil when the plant dies and decomposes. If the crops are harvested, the nitrogen lost when the organic matter is removed must be replaced by nitrogen fertilizers or legumes. Some possible nitrogen activities for a soil containing organic matter are shown in Figure 8, and some possible gains and losses are shown in Table 2.

Legumes

Legumes grown on minesoils are likely to increase the amount of soil nitrogen, as minesoils are low in organic matter and nitrogen. However, it has been found that legumes can reduce the total amount of soil nitrogen if the amount of nitrogen in the soil was high before the legumes were grown. This is especially true if the legumes are harvested and removed from the soil. Under these circumstances it seems reasonable to add as little nitrogen fertilizer as possible when growing legumes.

Legumes are members of a plant family in which practically all members benefit from a natural system that takes nitrogen gas from the air and changes it into a chemical form that can be used by plants. The change from nitrogen gas to useable nitrogen is called nitrogen fixation and is actually done by bacteria living in nodules on the legume root.

Table 3. Best Performing *Rhizobium* Bacteria for Each Legume[1]

Bacteria (scientific name)	Legume (generic name)	Legume (common name)
R. meliloti	*Medicago*	Alfalfa
R. meliloti	*Melilotus*	Sweet clover
R. meliloti	*Trigonella*	Fenugreek
R. trifolii	*Trifolium*	Clovers
R. leguminosarum	*Pisum*	Pea
R. leguminosarum	*Vicia*	Vetch
R. leguminosarum	*Lathyrus*	Sweetpea
R. leguminosarum	*Lens*	Lentil
R. phaseoli	*Phaseolus*	Beans
R. lupini	*Lupinus*	Lupines
R. lupini	*Ornithopus*	Serradella
R. japonicum	*Glycine*	Soybean
R. japonicum	*Vigna*	Cowpea
R. japonicum	*Lespedeza*	Lespedeza
R. japonicum	*Crotalaria*	Crotalaria
R. japonicum	*Pueraria*	Kudzu
R. japonicum	*Arachis*	Peanut
R. japonicum	*Phaseolus*	Lima bean
R. japonicum	*Alysicarpus*	Alyceclover
R. japonicum	*Stizolobium*	Velvetbeans

[1]Birdsfoot trefoil, crownvetch, black locust, sesbania and others require special bacteria not listed here.

The generic name of these bacteria is *Rhizobium,* and there are many species in this genus that form nodules on legume roots and fix atmospheric nitrogen. The legume uses the nitrogen fixed by the bacteria and in turn the legume provides the bacteria with certain nutrients needed by the bacteria. Many of the plants used in surface mine reclamation are legumes, such as clover, alfalfa, sweet clover, pea, bean, lespedeza, and vetch.

Different legumes need different species of bacteria to produce maximum amounts of useable nitrogen (Table 3). The bacteria needed by the legume planted on a particular soil may already be present in the soil. However, it is safer to inoculate seed with the proper bacteria before planting. These bacteria are commercially available and can usually be purchased from the same company that sells the legume seeds.

Legume seeds are inoculated by wetting them slightly with water, syrup, milk, or other adhesive. They should be wetted enough to make bacteria stick to the seeds, but not enough to make the seeds stick to each other. The inoculum, containing the bacteria, is sprinkled over the wet seeds and the two are thoroughly mixed. Inoculation should be done

Table 4. Nitrogen Content of Common Nitrogen-Containing Fertilizers

Chemical	N (percent)
Anhydrous ammonia (ammonia gas, NH_3)	82
Aqua ammonia (ammonia gas in water, NH_4OH)	24
Ammonium nitrate (NH_4NO_3)	33
Ammonium sulfate ($(NH_4)_2SO_4$)	21
Urea ($CO(NH_2)_2$)	45
Nitrogen solutions (ammonia gas, ammonium nitrate, and urea mixed individually or in combination with water)	19-37
Ammonium phosphate (monobasic) ($NH_4H_2PO_4$)	12
Ammonium phosphate (dibasic) ($(NH_4)_2HPO_4$)	18
Animal manure	10-20
Sewage sludge	5-10

immediately before sowing the seeds. This is best because the bacteria can be killed by drying or high temperatures. The soil should remain moist from the time of sowing until the seeds germinate in order to keep the bacteria alive until they can invade the seedling root. High soil temperatures will also kill the bacteria. These two possibilities are a good reason for applying a mulch in order to retain moisture at the soil surface and lower the surface soil temperature. More nitrogen is fixed by legumes if the soil is at a pH of 5.0 or higher. The bacteria will not penetrate many of the legume roots if the soil pH is below 5.0. A low soil pH is usually due to low soil calcium content, high sulfur content, or toxic amounts of aluminum. The solution to these problems is the addition of limestone to the acid soil.

Another cause of nitrogen fixation failure is the absence of molybdenum or cobalt from the soil. These elements are needed by the bacteria to convert nitrogen gas to a useable form. Minesoils are not likely to be deficient in these elements; but, if needed, the addition of small amounts of molybdenum or cobalt to a deficient soil can solve this problem. If legume roots have no nitrogen-fixing bacteria, the roots obtain available nitrogen from the soil in the same manner as nonleguminous plants.

Nitrogen Fertilizer

Nitrogen is nearly always added to minesoils in the initial revegetation operation. Because the soil does not retain nitrogen well, except in organic matter, fertilizers containing nitrogen must be added more often than fertilizers containing other nutrients. Nitrogen fertilizers are usually in the form of ammonium (NH_4), nitrate (NO_3), or urea ($CO(NH_2)_2$). Table 4 lists the more common nitrogen fertilizers.

Ammonia gas (NH_3) is 82 percent nitrogen by weight. This gas can be injected directly into the soil, or the gas can be dissolved in water before injection. A water solution of up to 30 percent (by weight) NH_3 can be made. This solution would contain 24 percent nitrogen. When this solution is injected into the soil, it raises the soil pH immediately. However, after a few days soil organisms usually convert the ammonia (NH_3) to nitrate (NO_3). Due to factors of soil chemistry, this conversion causes the pH of the soil to drop. Therefore, the ultimate effect of adding ammonia to a soil is to cause a pH decrease or an increase in soil acidity. If necessary, this undesirable effect can be offset by periodically measuring soil pH and adding limestone to raise the soil pH when needed. When fertilizers containing ammonia or urea are added to a soil with a high pH (above 8.0) there is a possibility that much of the nitrogen in the fertilizer will go back into the atmosphere as gas if the soil and fertilizer are not well mixed.

Urea is not an ammonium fertilizer when applied to the soil, but it changes to an ammonium form soon after application. Once the ammonium form is present it behaves as any other ammonium fertilizer. However, research has shown that a greater percent of the nitrogen from urea may be lost through volatilization than from other solid nitrogen fertilizers. Urea contains the highest amount of nitrogen of any solid nitrogen fertilizer. It is a useful form when the most nitrogen in the least weight is needed. Such a need arises when nitrogen fertilizers are applied from aircraft.

Nitrate fertilizers dissolve in soil water with ease and the nitrate nitrogen can be quickly absorbed by plant roots. However, this solubility means that sodium, potassium, and calcium nitrates are subject to loss from the soil through leaching and volatilization. These nitrates do not make the soil more acid; in fact, they tend to make the soil less acid.

Nitrogen fertilizers are needed more frequently than other fertilizers due to the rapid loss of nitrogen through leaching, volatilization, soil erosion, and plant uptake. Nitrogen can usually be added every 2 months to nonleguminous crops with good effect during the growing season when soil moisture is adequate.

Phosphorous Fertilizer

Phosphorous is another troublesome nutrient for reclamationists. Plants need relatively large amounts of phosphorous but the amount of available phosphorous in a soil is often low. Most soil phosphorous is usually present in forms that are relatively insoluble in water and, therefore, unavailable to plant roots. In acid soils insoluble iron and aluminum phosphate compounds are formed, and in alkaline soils insoluble calcium phosphate compounds are formed.

Phosphorous tends to be most available to plants at a pH of about 6.0. This is the pH at which soil phosphorous is least likely to form insoluble compounds with aluminum and iron or calcium.

An insufficient supply of phosphorous causes a stunting of plant growth. As the supply becomes less and less, the plant often shows a

Table 5. Phosphorous Content of Phosphorous Fertilizers

Fertilizer	P_2O_5 (percent)	P (percent)	Water soluble (percent)
Rock phosphate	27–38	12–18	0
Ordinary or normal superphosphate	18–20	9	85
Triple superphosphate	44	19	85
Concentrated superphosphate	52	23	85
Ammonium phosphate (monobasic)	48	21	92
Ammonium phosphate (dibasic)	46	18	90
Phosphoric acid	54	32	100

reddish or purplish coloration of the leaves. Older plants may also show a yellowish coloration. Plants recover about 10 to 30 percent of the phosphorous that is added to a soil in fertilizer. Very little phosphorous is lost by leaching, so most of the 70 to 90 percent of phosphorous that is not absorbed by plants forms insoluble iron, aluminum, or calcium compounds. This means that large amounts of phosphorous may accumulate in a soil that is repeatedly fertilized with phosphorous.

These relatively insoluble phosphorous compounds have a slight solubility in water. When the soil has accumulated a large amount of these compounds, they can release enough phosphorous to fertilize crops for a long period of time. No more phosphorous fertilizer should be added when such an accumulation occurs. It is possible in some soils for excessive amounts of phosphorous to combine with the available iron in the soil and cause an iron deficiency. These phosphorous accumulations can be detected by a soil testing laboratory.

There are several sources of fertilizer phosphorous that are available to reclamationists (Table 5). The most insoluble form is rock phosphate. This material is usually phosphate ore that has been crushed or ground to an acceptable fineness for field application on acid soils. About 5 pounds of phosphorous from rock phosphate are required to give the same plant yield as 1 pound of phosphorous from a more soluble fertilizer such as superphosphate. This means that about 3 pounds of rock phosphate would have to be applied for the same growth effect that 1 pound of superphosphate would provide.

Rock phosphate contains from 27 to 38 percent P_2O_5 (12 to 18 percent P), depending on the source of the ore. This P_2O_5 is traditionally called phosphoric acid by the fertilizer industry, although P_2O_5 is not the formula for phosphoric acid.

Superphosphate is also called ordinary or normal superphosphate. This phosphorous fertilizer contains about 20 percent phosphoric acid (P_2O_5)

Table 6. Potassium Content of Common Potassium-Containing Fertilizers

Fertilizer	**K_2O (percent)**
Potassium chloride (muriate of potash (KCl))	60
Potassium sulfate (K_2SO_4)	50
Potassium magnesium sulfate ($K_2SO_4 \cdot MgSO_4$)	22
Potassium nitrate (KNO_3)	44
Potassium carbonate (K_2CO_3)	57

or 9 percent phosphorous (P). About 85 percent of this phosphorous is soluble in water. Superphosphate is about 20 percent calcium and 10 percent sulfur, which is important for soils that are low in calcium or sulfur. It also contains traces of iron, copper, manganese, zinc, magnesium, and aluminum.

Superphosphate is manufactured by applying sulfuric acid to rock phosphate. Such treatment converts the insoluble calcium compounds of phosphorous to water-soluble compounds that can provide available phosphorous to plants.

Triple superphosphate contains about 44 percent P_2O_5 (19 percent P). Concentrated superphosphate contains about 52 percent P_2O_5 (23 percent P). Both these fertilizers are manufactured by treating rock phosphate with phosphoric acid. Neither of these fertilizers contains much sulfur, such as that found in ordinary superphosphate. This should be considered when fertilizing soils that are deficient in sulfur. The ammonium phosphates are important sources of phosphorous and will be discussed under mixed fertilizers. There are other phosphorous fertilizers, but the ones discussed here are those most likely to be encountered and most likely to appear in mixed fertilizers. When buying phosphorous fertilizer, it is wise to consider the pH of the soil, whether or not the soil may need sulfur, and the actual cost per pound of the available phosphorous in the fertilizer.

Potassium Fertilizer

Potassium is the third of the elements that are most likely to be deficient for plant growth in a soil. Soil potassium is usually considered to be present in three forms. Like phosphorous, most soil potassium (about 90 percent) is in a form that is unavailable to the plant. However, there is a form that is slowly available (up to 10 percent) to plants. Finally there is a readily available form (about 1 percent). Fortunately the slowly available potassium helps replenish the readily available potassium as the readily available is removed by plants or leaching. The unavailable form of potassium also becomes available with time, but this is a slow process and probably amounts to only a few pounds per acre per year.

Table 7. Analyses of Fertilizers Containing Calcium (Ca), Magnesium (Mg), and Sulfur (S)

Fertilizer	N	P_2O_5	K_2O	Ca	Mg	S
Ammonium nitrate limestone	20.5[1]			7.3	4.4	0.4
Superphosphate						
Normal or ordinary		18–20	0.2	20	0.2	12–14
Triple or concentrated		44–52	0.4	14	0.3	1.4
Calcium sulfate (gypsum)			0.5	22	0.4	17
Magnesium sulfate (epsom salts)				2.2	9.7	13
Ammonium sulfate	21			0.3		23.7
Potassium sulfate			50	0.7	1.2	17–18
Sulfur						50–99
Manganese sulfate						14–17
Ammonium phosphate-sulfate	15	20–39	0.2	0.3	0.1	15.4

[1]Numbers indicate percent.

When the soil does not contain enough available potassium, the older parts of the plant tend to lose their potassium because it is transported to younger and more actively growing tissues. Therefore, older leaves may lose their green color and the edges and tips of leaves may appear brown and dead. As the deficiency becomes more acute, the younger leaves may also show these symptoms.

There are several sources of fertilizer potassium (Table 6). Some of these sources are: muriate of potash or potassium chloride (KCl), potassium nitrate (KNO_3), sulfate of potash or potassium sulfate (K_2SO_4), potassium-magnesium sulfate ($K_2SO_4 \cdot MgSO_4$), and potassium carbonate (K_2CO_3). The most popular of these potassium fertilizers is muriate of potash or potassium chloride (KCl). Potassium chloride occurs in nature as large deposits of salt in various parts of the world. The salt is purified to a degree by removing such impurities as sodium chloride. Sodium is one of the elements that is undesirable in large amounts in soils to be used for plant growth. The potassium chloride in fertilizers is approximately 60 percent K_2O or 50 percent potassium (K). Potassium chloride is soluble in water and the readily available form tends to leach out of the root zone, particularly in sandy soils. Traces of calcium, magnesium, sodium, iron, sulfur, bromine, and boron are often found in fertilizer grade potassium chloride. Sandy soils in the eastern U.S. are likely to be deficient in potassium because potassium leaches easily and sandy soils do not readily retain nutrient elements.

Secondary and Micronutrient Fertilizer

The secondary nutrients are calcium (Ca), magnesium (Mg), and sulfur (S). These nutrients are used by plants in about the same quantities as

Table 8. Sources of Micronutrients

Chemical	Nutrient content
Zinc sulfate	Zinc (28)[1], sulfur (13), copper (0.02)
Zinc oxide	Zinc (77)
Copper sulfate	Copper (25), sulfur (13), zinc (0.5)
Copper oxide	Copper (75)
Iron sulfate (hydrated)	Iron (20)
White copperas (iron sulfate)	Iron (34)
Manganese sulfate	Manganese (25), sulfur (17), copper (0.5), zinc (0.08), boron (0.3)
Borax	Boron (12)
Polybor	Boron (20)

[1]Numbers in parentheses indicate percent.

phosphorous. However, nitrogen and potassium are used in larger quantities. Soils may be deficient in these secondary nutrients, but not nearly as often as they are deficient in one of the primary nutrients. Also, many of the fertilizers that provide primary nutrients also provide secondary nutrients. For instance, superphosphate provides large amounts of calcium and sulfur, and a small amount of magnesium, along with the phosphorous.

Table 7 shows some of the chemical materials that are often used to supply secondary nutrients. If the soil testing laboratory recommends that a secondary nutrient be applied, it is usually one of the materials shown in this table. If the laboratory does not specify a material, the reclamationist can use the percent values in Table 7 to calculate the amount of material needed to satisfy recommendations.

Some nutrients are needed in even lesser amounts than the secondary nutrients. These are termed micronutrients. They are vital for plant growth, but they are only needed in very small amounts and are usually present in the soil or as contaminants in macronutrient fertilizers. Table 8 shows the chemical materials that are often used to supply micronutrients for deficient soil.

Mixed Fertilizers

A multinutrient or mixed fertilizer is one that contains at least two major plant nutrients. These fertilizers can be made by manufacturing individual granules that contain all the nutrients needed in the fertilizer. The multinutrient fertilizer can also be made by blending individual nutrient materials where each granule contains only one of the needed nutrients. For instance, a blended fertilizer could be a mixture of individual ammonium nitrate, superphosphate, and potassium chloride granules.

The amounts of nutrients in a fertilizer are given by a three-number designation such as 10-20-15. The first number indicates that the fertilizer

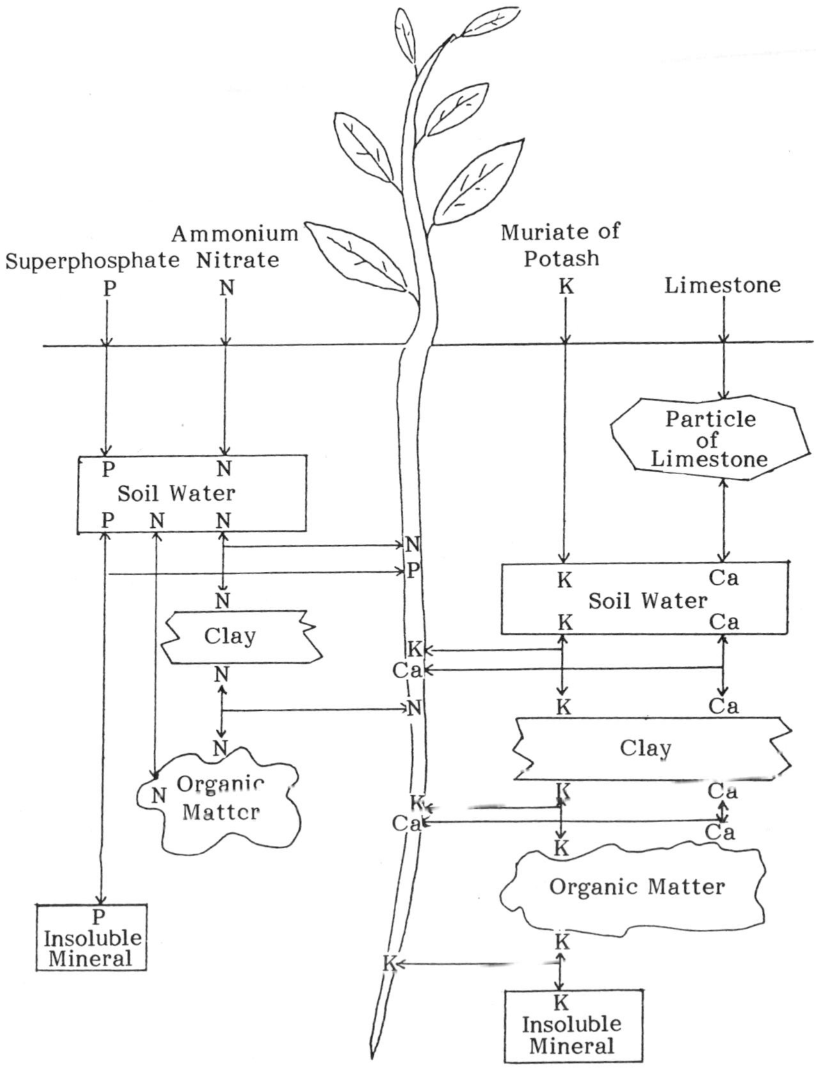

Figure 9. Possible pathways taken by nutrient elements when applied to a soil.

contains 10 percent nitrogen. The second and third numbers indicate that the fertilizer contains 20 percent P_2O_5 and 15 percent K_2O, respectively.

The values for P_2O_5 can be converted to values for P by multiplying P_2O_5 values by 0.437. Values for K_2O can be converted to values for K by multiplying K_2O by 0.83.

Two of the more popular multinutrient fertilizers are monoammonium phosphate ($NH_4H_2PO_4$) and diammonium phosphate ($(NH_4)_2H_2PO_4$). Both these fertilizers are solid and do not absorb moisture easily. Monoammonium phosphate contains 11 percent nitrogen and 48 percent P_2O_5 (21 percent phosphorous). Diammonium phosphate contains 16 to 21 percent nitrogen and 46 to 53 percent P_2O_5 (20 to 26 percent phosphorous), depending on how it is manufactured. Both of these fertilizers are used in mixed fertilizers.

The fate of the fertilizer elements added to a soil is complicated. However, many of the pathways taken by these elements when in the soil are known. Figure 9 describes some of the possible pathways that nitrogen, phosphorous, potassium, and calcium might take on their way to the plant root.

SOIL ACIDITY

Soil acidity, soil calcium, soil magnesium, and liming are all closely related where soils and plant growth are concerned.

Acid soils occur naturally in most of the eastern half of the U.S. However, surface mining is considered to be a prime source of acid soils by many people. This is true in many instances, but surface mining also produces alkaline soils, the opposite of acid soils. Soil acidity causes problems that result in poor plant growth. These problems can be grouped into three general categories: decreased availability of nutrients, increased availability of toxic elements, and decreased activity of beneficial soil organisms. Figure 10 shows the general effect of soil acidity on nutrient availability and microbiologic activity.

Soil organisms such as bacteria and actinomycetes are adversely affected by soil acidity. When the soil pH decreases to approximately 5.5, both bacteria and actinomycetes activity decreases significantly. This is important for such things as nitrogen fixation by legumes and other nitrogen-fixing organisms. Fungi are not greatly affected by soil pH itself, but fungi become less active at a higher pH due to the increased activity of bacteria and actinomycetes and the resulting competition.

The alkalinity or acidity of a soil is measured by using a scale called pH. This scale ranges from 0 to 14. Most natural soils range in pH from 4.0 to 8.5. The value 7.0 is neither acid nor alkaline and is considered neutral. All pH values above 7.0 are technically alkaline. However, for practical purposes, any pH between 6.0 and 8.0 is considered near neutral. Most crops grow best at a pH between 6.0 and 7.0. Some plants, such as azaleas, Irish potatoes, and cranberries, grow best on a more acid soil. Keep in mind that a pH unit may only increase by one unit but the increase in strength is 10 times. For instance, soil with a pH of 5.0 is 10 times as acid as a soil with a pH of 6.0, and a soil with a pH of 4.0 is 100 times as acid as a soil with a pH of 6.0. A soil with a pH of 9.0 is 10 times as alkaline as a soil with a pH of 8.0.

The usual method of correcting soil acidity is to add finely ground limestone and disc it into the soil. The amount of limestone needed to counteract soil acidity is determined by several factors other than soil pH, and the amount needed to raise pH by one unit can vary from soil to

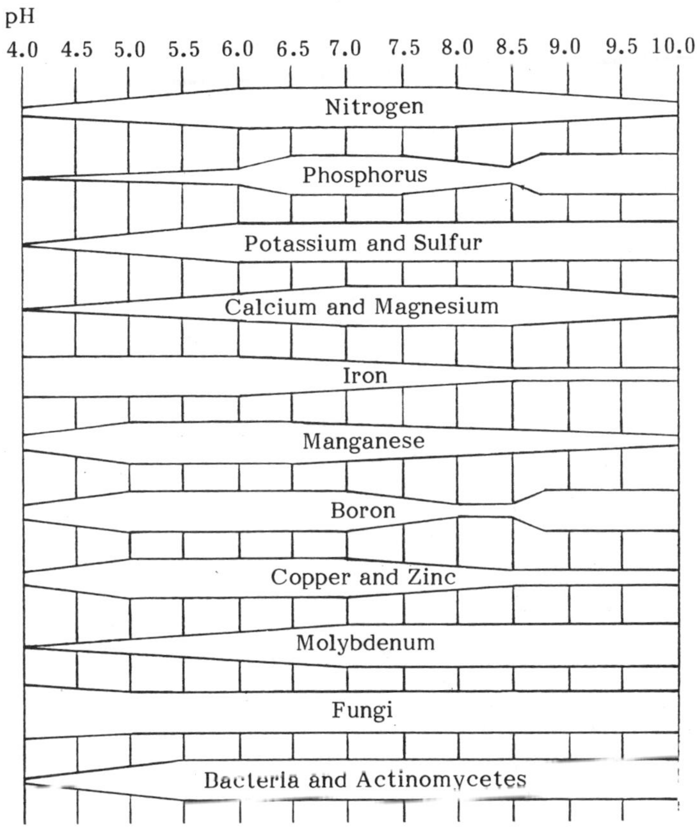

Figure 10. Effect of soil acidity on nutrient availability and microbiologic activity. Maximum availability and activity are indicated by the widest part of the bar.

soil. Soil testing laboratories can determine the amount of lime needed for different soils.

Pure lime is technically calcium oxide (CaO). However, the term lime is used to mean other materials such as calcium hydroxide ($Ca(OH)_2$) and calcium carbonate ($CaCO_3$). Calcium carbonate in the form of ground limestone is the material most often used to raise the pH of soils. The fineness to which the limestone is ground is of prime importance. In order to quickly neutralize soil acidity, limestone should be fine enough so that about 90 percent will pass through a 10-mesh screen and 50 percent through a 60-mesh screen. Many states have laws governing the fineness of agricultural grade limestone. Limestone that will pass through a 100-mesh screen can raise soil pH from 5 to approximately 6.5 in 2 weeks. However, 25-mesh limestone would only raise the pH from 5.0 to approximately 5.2 in 2 weeks.

Limestone must be well mixed with the soil in order to be most effective. Mixing the limestone with the soil to a depth of at least 6 inches is recommended.

Agricultural grade limestone has several beneficial effects on soils. It corrects soil acidity, supplies calcium to the soil, supplies magnesium to the soil (if dolomitic limestone is used), increases the availability of phosphorous, improves the physical properties of soils, reduces the activity of injurious agents in the soil, increases the speed of organic matter decay to release nutrients to the soil, and increases the crop yield through all of these actions.

There are two general types of limestone, calcitic limestone ($CaCO_3$) and dolomitic limestone ($CaMg(CO_3)_2$). Dolomitic limestone is sometimes preferable (where magnesium is deficient) because it neutralizes acidity as well as calcitic limestone and it also adds magnesium to the soil for plant use. However, minesoils often have high magnesium contents and the addition of dolomitic limestone may cause excessive soil magnesium and revegetation problems. Under this circumstance, it is probably best to add dolomitic limestone only when the minesoil is low in magnesium. Calcitic limestone contains about 85 to 95 percent calcium carbonate ($CaCO_3$) and dolomitic limestone contains about 52 percent calcium carbonate and 42 percent magnesium carbonate ($MgCO_3$).

The mixture of soil and stone material that is removed from above the coal seam is somewhat different from the topsoil that is replaced over this mixture. This mixture of overburden can have both acid and alkaline elements at the same time. Some of the original geologic layers may have been acid, whereas others were alkaline. This means that the overburden mixture may become acid or alkaline after long exposure to weather, depending on the relative amounts of acid and alkaline material in the overburden mixture.

The amounts of acid- and alkaline-producing materials in an overburden can be estimated with laboratory procedures. If it is found from these estimates that the acid materials will eventually predominate and form an acid soil, limestone can be added as needed to control this acid soil formation. If alkaline materials are found to be dominant, there is usually no attempt to control the alkalinity; instead, plants that can tolerate high pH are planted. These acid and alkaline determinations can be made for the soil and rock strata prior to the mining operation and can be used to predict the acidity or alkalinity of the resulting overburden.

CHAPTER 2

TILLAGE

Tillage is the operation of plows or other implements to prepare the soil for the germination of seeds and the growth of plants. Ordinarily there are three purposes of tillage: (1) to prepare a suitable seed germination and root growth medium; (2) to control unwanted plant growth; and (3) to improve the physical condition of the soil for rain infiltration, water retention, aeration, and minimal erosion. Surface mine reclamationists are not usually concerned with unwanted plants, but they are most concerned with the other two purposes of tillage.

SOIL WATER

The best tillage for surface mine reclamation is one that leaves a seedbed that is loose enough to allow water infiltration but compact enough to form micropores for water movement to the germinating seeds. Also, tillage should be done on a contour to minimize soil erosion until the planted crop becomes an effective erosion control agent. Finally, the least amount of tillage that can be used to obtain objectives is usually best. When soil is first loosened by tillage it expands and has more pore space for water and air storage. However, repeated tillage can pulverize the soil and reduce the pore space and water storage capacity.

The moisture content of the soil determines the time of tillage to a large extent. If the soil is too dry when tilled it tends to lose any structure that it may have and falls apart into individual soil particles. This may reduce the number and size of soil pores and restrict water movement and storage. If the soil is too wet, tillage causes dense clods to form in silty or clayey soils. This can also reduce soil porosity. Plowing a wet sandy soil probably does not do as much damage as plowing a wet soil of fine or clayey texture. In fact, plowing a moist sandy soil has been found to pack some of the sandy soil particles into loose clods that improve soil structure.

Table 9. Approximate Bulk Density and Available Water Found for Soils of Different Textures[1]

Soil texture	Field capacity (percent)	Permanent wilting (percent)	Bulk density (g/cc)	Available water (in./in. soil depth)
Silt loam and clay loam	26.0	7.0	1.30	.26
Loam	25.0	6.5	1.39	.25
Very fine sandy loam	22.0	5.5	1.35	.22
Fine sandy loam	20.0	5.5	1.35	.20
Sandy loam	16.0	4.0	1.40	.17
Loamy fine sand	13.5	3.5	1.40	.14
Loamy sand	11.0	3.0	1.45	.12
Fine sand	9.0	3.0	1.45	.10
Sand	9.0	2.0	1.50	.09
Coarse sand	6.0	2.0	1.60	.06

[1]These values should be used as estimates only. Actual values for any specific soil could vary significantly.

SOIL COMPACTION

It has been known for many years that any attempt to regrade surface mines results in soil compaction. This compaction is one of the prices that must be paid for the return of a rough landscape to a more useful and aesthetically pleasing one. Soil compaction can occur from natural causes or from human activities. The surface mine reclamationist is mostly concerned with the compaction caused by humans through topsoil removal, overburden regrading, and topsoil replacement. The process of topsoil removal destroys much of a soil's structure. If the soil is placed in a stockpile, more of the structure will be lost as the soil-carrying vehicles pass over the stockpile repeatedly. Structure is further destroyed when the vehicles remove the soil from the stockpile. Finally, compaction takes place as vehicles place the topsoil on the reclaimed area and the final regrading is done.

In compacted soils the soil particles are packed closely together and few macropores occur. In fact, soil compaction may be thought of as a loss of pore space. This loss of pores means that less water can move from the soil surface into the soil. There may also be less water storage space in the soil and less space for air movement into the soil. Finally, a compacted soil is harder for roots to penetrate than a loose soil with structure.

Wet soils compact more easily than dry soils because water acts as a lubricant between soil particles and any pressure on the wet particles allows them to slide over each other until all their surfaces are in close contact, leaving little open space to form soil pores.

As a soil becomes compacted the bulk density of the soil increases. Bulk density is the weight of a given volume of soil and is usually expressed as grams per cubic centimeter of soil. A soil with a low bulk density will be light in weight, loose, and have many pores. A soil with a high bulk density will be heavy, compact, and, therefore, will have less pore space. Bulk density can be considered a measure of pore space in the soil.

The usual method of determining bulk density is to drive an open-ended metal cylinder of known volume into the soil, remove the cylinder with soil, and dry the amount of soil that exactly fills the cylinder. Weigh the dry soil and divide the weight by the known volume of cylinder.

Normal root growth is restricted as the bulk density becomes greater than 1.5 g/cc in fine-textured (clay and silt) soils. In coarse-textured (sandy) soils, bulk densities above 1.7 g/cc usually restrict root growth. Table 9 shows typical bulk densities and available water for several soil textures.

Compaction is caused by destroying soil structure in some manner. Structure can be destroyed by moving a soil in some way, such as tilling or lifting topsoil for replacement. It can also be destroyed when heavy equipment or animals pass over the surface repeatedly. All of the above actions are intensified if the soil is moist enough to lubricate soil particles and help them slide over each other when subjected to pressure. Rainfall on bare soil can also cause compaction by destroying structure and washing the finer soil particles into soil pores.

Compaction can be overcome by the growth of plant roots through the soil, causing channels to form when the roots die and decay. Organic matter from dead roots and tops adds material to the soil that promotes structure by causing soil particles to aggregate into small, loose, and porous lumps. Because of their organic matter production and strong root penetration, perennial grasses and legumes are among the best crops for promoting the factors that reduce the effects of compaction.

Almost any type of plowing, including subsoiling, will temporarily reduce compaction. When soil is first loosened by plowing, it tends to expand; ie, more open or pore space is created between the clods of soil formed by the tillage action. Tillage is repeated until the desired clod size is obtained. If the plowing is repeated beyond this point, the clods will become smaller and smaller until they begin to fill the original spaces between clods. Therefore, a single or few plowings may increase a soil's ability to accept and hold water, but repeated plowing will probably cause the soil to compact and prevent rainwater from moving into the soil. Instead the water either stands on the soil surface or flows over the soil surface, causing erosion. In either case the soil is deprived of water that will be needed for seed germination and plant growth.

Plowing of some type is needed to prepare a seedbed. In the case of surface mine reclamation, it is also often needed to loosen compacted soils. It is likely that both topsoil and overburden will be compacted. Therefore, the deepest practical cultivation will probably be most beneficial for water storage and plant growth.

SUBSOILING

Deep tillage is usually referred to as subsoiling. There are several different types of subsoiling implements. The best implements are probably ones that penetrate the soil to the desired depth and then lift the soil, producing cracks for water movement, air movement, and root penetration. For the cracks to remain open for a long period, subsoiling should be done when the soil is moist or dry, but not wet. If the soil can be molded into a clod by applying moderate pressure with the fingers, the soil is probably too moist. Wet soil is likely to flow back into the cracks after the subsoiler has passed on. If strong pressure is required to form a clod, or if soil feels moist but will not form a clod, the soil is probably at the right moisture content for subsoiling. Dry soil may form cracks as the subsoiler moves through the soil, but some soils will become loose enough to fall back into the cracks and close them after the subsoiler passes. However, some reclamationists have found that soils in their areas respond best to subsoiling when dry.

SANDY SOILS

Coarse-textured or sandy soils are likely to be loose and soft. They will probably have good aeration, good rain infiltration, good drainage, and easy cultivation. These soils do not hold moisture and nutrients as well as finer soils and are likely to be poor soils for plant growth. Most of the problems with sandy soils can be helped by adding organic matter to the soils. The reclamationist often adds organic matter to minesoils with straw mulches and by growing perennial grasses and legumes. Any decomposed organic matter will help bind the sand grains together and hold moisture for plant use. Organic matter will also act as a nutrient storehouse and increase the fertility of the sandy soil.

SILTY OR CLAYEY SOILS

Fine-textured (silty or clayey) soils are different from sandy soils in that they tend to stick together too well. This means that when they do not have a structure they have a tendency to be compact and impenetrable to plant roots and water. When water does penetrate fine-textured soils, they may hold the water so tightly that plants cannot obtain much of the water.

If these soils are plowed when wet, they tend to lose any structure and porosity they may have. If plowed when dry, they form hard clods that produce a poor seedbed. These soils should be tilled at the moisture content previously described as best for subsoiling.

The other method of obtaining and maintaining structure in these soils is to increase the organic matter content. Organic matter causes the solid mass of an unstructured clay soil to form small separate clods with space between the clods for water, air, and root movement. Fortunately the principal plants of reclamation, legumes and perennial grasses, are some of the best plants for adding organic matter to soils and promoting structure. These forage plants do not require as much cultivation as row crops, and this tends to prevent organic matter loss and reduce the amount of soil compaction caused by wheel traffic and implement cultivation.

CHAPTER 3

SOIL TESTING, INTERPRETATION, AND RECOMMENDATIONS

The purposes of soil testing are to determine the nutrient status of a soil, the pH of a soil, and the amount of fertilizer and lime needed by the reclamationist to produce economically a soil cover of vegetation.

The process of determining the nutrients in a minesoil and the kinds and amounts of amendments to be added is usually referred to as soil testing. Soil testing is actually a series of operations. First, a sample of the soil to be tested must be obtained. Second, the soil pH and lime requirements are determined. Third, the nutrients must be extracted from the soil in order to determine which nutrients are present and in what quantities. Fourth, the kinds and amounts of nutrients present in the soil extract are determined by chemical analyses. Fifth, the results from the chemical analyses are interpreted in terms that can be used to make recommendations for the types and amounts of lime and fertilizer needed. Sixth, the fertilizer and liming recommendations are made to the reclamationist.

SOIL SAMPLING

The soil sample that the reclamationist sends to the soil testing laboratory must be representative of the area to be vegetated. This means that the reclamationist must determine which areas are uniform insofar as physical appearance is concerned. Of course, the reclamationist cannot know the chemical differences, but differences in color and texture should be noted. Areas that have obvious differences should be sampled separately.

When an area has been selected for sampling, 20 subsamples should be taken in the area. These subsamples can be taken in a random fashion that covers the whole area, or the area can be divided into approximately 20 blocks of equal size and a subsample taken from each block. If small areas within the subsample area appear to be different from the rest of the subsample area, the different area should be avoided. These small and different areas may not respond well to the subsequent lime and fertilizer recommendation, but the cost of sampling and testing all such small areas would probably be prohibitive. The subsamples are placed in a container and mixed thoroughly. After mixing, approximately 1 pint or 1 pound of this mixture is sent to the testing laboratory as a composite sample.

The method by which subsamples are taken is important. Each subsample should be of the same size as all other subsamples. This is usually accomplished by pushing a soil sample tube into the soil to a predetermined depth. The depth should be 9 inches unless the replaced topsoil is less than 9 inches. In that case the topsoil should be sampled down to the overburden. Other instruments, such as a shovel or trowel, can be used but care should be taken to obtain subsamples of approximately equal size that represent the entire depth. This 9-inch depth does not represent all the soil volume that will be used by the plant root, but it does give a standard volume of soil to be used for soil amendment recommendations.

SOIL NUTRIENT EXTRACTION

When the composite soil sample reaches the soil testing laboratory, the nutrients must be extracted from the soil to determine the kinds and amounts of nutrients in the soil. This is usually done by leaching the soil with a mild acid solution. The amounts of nutrients extracted may not be the same amounts available to plants, but they will be amounts that can be correlated with plant growth. Different laboratories use different extracting solutions, and the same laboratory may use different extracting solutions for different soils. The important thing is that the laboratory must be able to correlate the amount of nutrient found in the soil with the amount of plant growth that the nutrient will support.

SOIL EXTRACT ANALYSIS

After extraction, the laboratory personnel will determine the nutrients and their amounts in the soil extract. This part of the soil testing procedure is easily reproducible and is probably the simplest and most error-free part of soil testing.

The reclamationist usually needs information regarding soil pH, nitrogen, phosphorous, and potassium. Laboratory tests can be made for all of these plus many more, such as sodium, calcium, magnesium, boron, arsenic, sulfur, zinc, copper, molybdenum, iron, and aluminum. Soil tests for nitrogen have not been as useful as tests for pH, phosphorous, and potassium. Therefore, many laboratories do not test for nitrogen unless a specific request is made. Available forms of nitrogen remain in the soil for short periods of time. Therefore, it seems more reasonable to assume that any minesoil will need nitrogen and make a recommendation based on the needs of the plant to be grown.

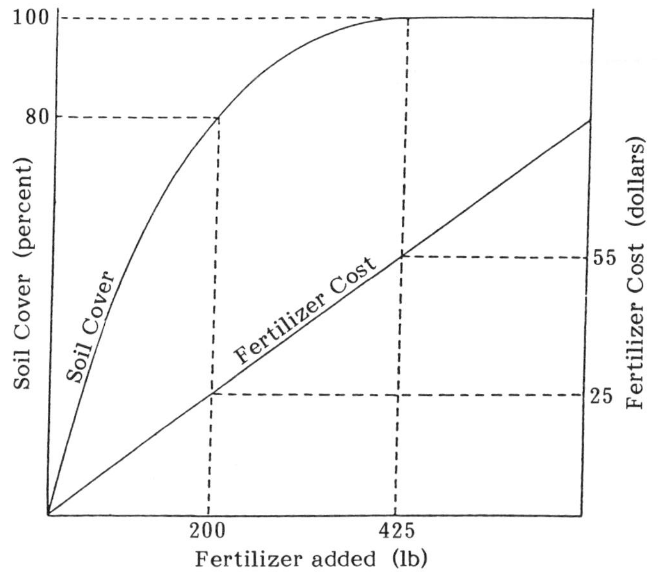

Figure 11. Relationship between amount of fertilizer applied and soil cover obtained.

INTERPRETATION

The results from the laboratory must be interpreted in such a way that the amount of fertilizer needed to successfully grow the desired crop will be recommended. This information is gathered over a period of time by growing a specific crop at different levels of fertilization on a specific soil and determining which level gives the desired crop yield. Once this information plus the soil test information is known, the soil scientist can predict the amount of fertilizer needed to obtain a desired crop yield. It should be remembered that many other factors also affect crop yields, such as soil moisture, insects, disease, temperature, seed quality, and skill of the reclamationist. Soil nutrients are only one factor to consider in crop yield or soil cover.

RECOMMENDATIONS

Finally the lime and fertilizer recommendations can be made. Recommendations may be based on several factors such as present soil fertility, crop to be grown, type of planting, amount of plant growth desired, and method of fertilizer application. However, most recommendations are based only on present soil fertility and a reasonable yield that can be obtained at the least cost. This reasonable yield is usually determined by the value of the crop. Increasing amounts of fertilizer are added until the increase in yield obtained will not return a profit over the cost of the increase in fertilizer. This point is difficult to determine in reclamation because its usual objective is to obtain enough ground cover for erosion control rather than production of a saleable crop. Therefore the optimum fertilizer rate for reclamation may be less than that for crop production. Figure 11 shows the relative relationship between the cost of fertilization and the amount of soil cover obtained. This figure indicates that there is likely to be a large difference in the costs of fertilizers needed for 80 and 100 percent ground covers. Whereas it may only cost $25 to obtain 80 percent soil cover, the cost more than doubles to obtain another 20 percent to reach 100 percent soil cover. The values in this figure are for illustration purposes only.

CHAPTER 4

SEEDS

A mature seed contains a miniature inactive plant called the embryo. When factors such as moisture, temperature, oxygen, and light are present within acceptable limits, the embryo becomes active, begins to develop, and produces a growing plant. This process is called seed germination and continues until the new plant can manufacture all the food it needs without using food that had been previously stored in the seed.

Moisture and oxygen are the two most important factors for germination of all plant seeds. Some seeds also have a narrow temperature range for germination. The seeds of some plants can germinate at 3°F, whereas others can germinate at 120°F. However, most seeds germinate well at a temperature of about 70°F. Another factor that has some effect on certain seeds is light. Light is necessary for germination of some seeds, and inhibits germination in others, but most seeds can germinate in light or darkness. Many of the grasses, such as Bermuda grass and Kentucky bluegrass, germinate more quickly when they are exposed to light. It would be wise to know the light requirements before planting any unfamiliar seeds, but practically all seeds used by reclamationists will germinate in light or darkness.

MOISTURE

For successful seed germination the soil should have about 50 to 75 percent of the total amount of water that it can hold. A saturated soil (100 percent) might prevent the seed from obtaining the oxygen needed for germination. Seeds must absorb between 25 and 75 percent of their weight in water before they start to germinate. Millet absorbs about 25 percent before starting germination, whereas soybeans must absorb about 75 percent before starting to germinate. The amount and rate at which water is absorbed by a seed depends on: (1) the amount of water in the soil surrounding the seed; (2) the compactness of the soil surrounding the seed; (3) the degree of contact between the soil and the seed; (4) the temperature of the soil; and (5) characteristics of the seed.

If the moisture content of the soil is below about 50 percent of its capacity, the rate of seed germination will be reduced. When the soil moisture content reaches the point at which plants wilt, seeds cannot absorb enough water to germinate, even at a slow rate.

If the soil is packed too loosely around the seeds, there is not enough contact between the seeds and the moist soil for the seeds to absorb adequate moisture for germination. If the soil is packed too tightly, the seeds may not receive enough oxygen to germinate. Therefore, the planter should bring the seeds in close contact with the soil but leave enough soil pore space for air circulation.

Most grasses and legumes that grow in the warm season germinate well at about 90°F. Cool season crops germinate well at about 70°F. When seed testing laboratories test seeds for germination ability, they use temperatures of 68°F at night and 86°F during the day. It has been found that this alternating of temperatures favors germination of the seeds of many plant species.

SEED DORMANCY

Not all seeds will germinate when exposed to the conditions that are usually considered favorable for germination. These seeds are called dormant seeds and must be specially treated before they will germinate. Some dormancy is caused by hard or thick seed coats that will not absorb moisture. This dormancy can sometimes be broken by cutting through the thick coat, or by some combination of freezing and thawing the seeds. Another type of dormancy is caused by the chemical makeup of the seeds; this can be broken by storing the seeds at a low temperature under moist conditions.

SEED LAWS

Federal law requires that all seeds sold in the U.S. be labeled for variety, purity, and germination. Also, states have laws to regulate the quality of seeds sold within the state. These state laws usually require that an accepted name for the seed be shown on the label along with the seed purity, amount of weed seeds present, names of weed seeds, germination percent for the desired seed, and when the seed was tested. Seed purity is a measure of unwanted seeds, trash, or any other impurity found with the desired seeds. Percent germination is the percent of the desired seed that will germinate in a reasonable period of time.

CHAPTER 5

MULCH

Mulches have many effects on plants and soils. Mulches can be anything from straw, to shredded paper, to films of petroleum products sprayed on the soil surface, to sheets of plastic, to chicken house litter, to stubble left on a field after haying. The most common mulches for a reclamationist are straw, hay, and cellulose mulches (Figures 12 and 13). Straw and hay are usually applied when seeds are sown dry, and cellulose fiber is applied when seeds are sown in a water slurry.

Figure 12. Straw or hay mulch.

Figure 13. Cellulose mulch.

Surface mulches are important because they usually improve conditions at the surface of the soil. Conditions improved by mulches include: water infiltration, temperature, erosion control, seed germination, prevention of soil crust formation, soil structure formation, and nutrient supply. These factors are discussed under soil organic matter.

Other mulches that may be used are sawdust, tree bark, wood chips, woven materials, chemical fibers applied as a spray, plastic sheets, and shredded paper. Sawdust, tree bark, wood chips, and shredded paper are relatively cheap but are no better than straw and cannot be applied in a slurry as can cellulose fiber. Hay usually introduces seeds other than those planted. This is a disadvantage if a pure stand is desired, but may be an advantage if the principal aim is to get a soil cover as quickly as possible. The other mulches (sprays, plastic sheets, woven materials) are expensive but may have application in special situations such as water diversion ditch channels, steep soil banks, and dams.

Straw and hay mulches are usually applied at the rate of 1 to 2 ton/A. Bark, wood chips, and sawdust are applied at approximately 30 yd^3/A. The sprayable chemical mulches are applied according to the instructions that come with each product. Mulches that come as sheets give a continuous cover over the area to be mulched and should probably be chosen on the basis of lowest cost of material and application. The most popular thicknesses for polyethylene plastic sheeting are probably 4 mil and 6 mil. Plastic sheets or any other material that prevents water infiltration into the soil should be avoided if vegetation establishment is intended.

Figure 14. Straw mulch spread by mulch blower.

Hay or straw mulch can be applied by hand to small or inaccessible areas, and an asphalt emulsion can then be sprayed over the mulch to hold it in place. String or twine can also be stretched over the mulch and secured by stakes to hold the mulch in place. A less expensive method is to roll a crimper or flat discs over the straw to push parts of the straw into the soil.

The uniformity of any mulch application is most important. If an application is not uniform, the thick applications can retard or prevent seed germination or establishment, whereas the thin applications may provide none of the advantages of mulching. A thick mulch can prevent or retard seedling establishment in two ways: It can act as a physical barrier to seedling emergence, or it can prevent the soil from reaching a temperature high enough for seed germination. The temperature factor could be particularly important in the fall when the soil cools at night and does not warm during the day due to the shading effect of a thick mulch. Also, germination could be slowed in the spring if the soil surface was kept too cool by thick mulching. Nonuniform application can be caused by many factors such as defective equipment, unskilled operators, poor mulching material, or poor weather conditions.

One other possible problem with mulches is not completely understood. Under some conditions mulches seem to produce a toxic substance that kills young seedlings. This is probably more likely to happen where the straw is too thick, causing the soil surface to be poorly aerated.

An efficient method of applying mulch is with a mulch blower (Figure 14). These machines accept rectangular bales of hay, chop the plant stems, and blow them over a distance as great as 70 feet in calm weather. Some mulch blowers are equipped with asphalt emulsion injectors that apply asphalt as the mulch is blown, causing the straw stems to adhere to each other as they fall to the minesoil surface. Asphalt is applied at rates of 200 to 800 gal/A. If the hay or straw is being applied

Figure 15. Cellulose mulch, seeds, and fertilizer applied to minesoil by hydraulic seeder.

to a rolling or relatively level area, it can be held in place by passing an implement such as a disc or crimper over the mulch to press parts of the stems into the soil. Asphalt may be necessary on areas too steep for tractor operation. There are also mulch spreaders that accept round hay bales. Asphalt or a crimper would be applied in a separate operation. It has been found that a crimper can actually help to conserve soil moisture. Apparently this occurs because some of the crimped straw is left in a position perpendicular to the soil surface. Straw in this position protects the soil surface from drying winds and prevents soil water loss.

Cellulose fiber can be applied with a mulch blower, but it is more efficient to apply this fiber in a water slurry with a hydraulic seeder (Figure 15). For best results the fiber should be applied uniformly over the minesoil. The approximate rates of application are 1 ton/A for slopes and 0.5 ton/A for level acres. On calm days the slurry can be sprayed over distances up to 200 feet.

CHAPTER 6

REFORESTATION

In the eastern U.S. the establishment of trees on surface mines has been a traditional method of revegetating for minesoil stabilization and economic rehabilitation. Reforestation has lost much of its importance because of changes in reclamation regulations. Much of the land that is surface mined in some regions of the U.S. is forest land, and in most cases this land should be returned to forest in order to promote the best interests

Figure 16. Forest artificially established on minesoil.

Figure 17. Building constructed on minesoil.

of the landowner and the surrounding community (Figure 16). However, there can be no objection to returning previous forest land to some other use such as pasture, row crop, orchard, or building site if such use is in the best interests of the landowner and the community (Figure 17).

Most of the early tree planting on mines was done on overburden that had not been topsoiled, regraded, or vegetated with any other types of plants. If there were no extreme nutrient deficiencies, soil toxicities, or lack of soil moisture, most of these plantings were successful. However some states soon became concerned about the environmental problems associated with surface mining and they enacted laws requiring the regrading of overburden before revegetation. Regrading does result in a more pleasing and traversable landscape. However, it also causes minesoil compaction and this compaction has proved detrimental to tree and other plant growth.

An apparent problem with reforestation is the lack of control of soil erosion and sedimentation during early tree growth (Figure 18). It usually takes trees from 5 to 10 years to develop enough crown diameter and deposit enough organic matter to protect minesoils from the erosive effects of raindrops and flowing water. This problem resulted in governmental regulations that require quick and complete vegetative cover of minesoils after regrading. However, quick covers such as herbaceous legumes and grasses can be detrimental to tree survival and growth. Legumes and grasses often offer too much competition to tree seedlings for the available soil moisture, nutrients, and light.

A final obstacle to the tree planting is economic. Most mining companies would like to establish an accepted vegetative cover as quickly as possible in order to obtain bond release. An accepted cover is more likely to be obtained with the least effort and expense by planting grasses and legumes without the added problem of establishing trees on the same area. However, it would be wise for landowners and mine

Figure 18. Erosion on minesoil planted with pine.

operators to consider the possible long-term economic advantages of surface mine reforestation, along with the short-term economic gains from unused grasses and legumes.

DIRECT SEEDING

Direct seeding for purposes of reforestation has been a successful and useful technique with some forest species (Figure 19). It is usually more economical than seedling planting and can be used on steep slopes or rough terrain where seedling planting would be impossible. Also, a larger area can be planted in a shorter period with direct seeding, and much less labor is required. Regraded surface mines offer a special incentive for direct seeding because they have the bare mineral soil surface that is needed for tree seedling establishment. This incentive disappears when a forage stand becomes established. Direct seeding is the method used by nature to perpetuate forests. However, nature can afford to use great quantities of seeds and reseed a number of times if the first seedings are not successful. This is not true for humans, who want success with the least amount of seeds and a single application.

Two disadvantages of direct seeding are a lack of control over stocking and spacing. The spacing problem may be partially overcome by using row seeders pulled by tractors. Also, the stocking problem will probably be overcome as seeding rates are determined more precisely through experimentation and experience. The most serious disadvantage of direct seeding on surface mines is the fact that grasses and legumes have to be planted at the same time or before the tree seeds, and these plants usually offer too much competition to newly germinated tree

Figure 19. Tree stand produced by sowing longleaf pine seed on a minesoil.

seedlings. This disadvantage may be overcome by sowing the tree seeds in strips or spots where forage seeds have not been sown. Another possibility for overcoming the competition of a complete forage stand is to eliminate the forage in spots or strips by mechanical scalping or chemical deadening of vegetated areas within the forage stand.

One of the most significant factors affecting the success of direct seeding is the same as for any other vegetation success. This factor is soil moisture. There must be enough water in the minesoil for tree seed germination and establishment of the resulting seedling.

It has been found in the southern U.S. that fall and spring are the best times for direct seeding southern pines. Fall has usually, but not always, proved best for longleaf pine. Spring is usually best for loblolly, shortleaf, Virginia, slash, and eastern white pines. These two seasons are best because soil moisture is most likely to be near optimum for seed germination and seedling establishment during these periods in this region. Late June is considered the best time to sow pine seeds in the southwestern U.S. At that time the soil is usually moist enough for seed germination and seedling establishment. At other times the soil is either too dry or the temperature is too low for germination. The optimum time for sowing in each mining area must be carefully selected by the reclamationist based on soil moisture conditions and temperature.

Site Selection

Almost any area that can be planted with tree seedlings can also be direct seeded with tree seeds. There are at least two exceptions. A soil that tends to be droughty at the surface, such as a sandy soil, should be planted with seedlings because moisture is needed at the soil surface for seed germination. This problem could be partially overcome by mulching the soil surface with straw after sowing, or by covering the seeds with soil. The other poor site for direct seeding is a slope where there is nothing to hold the seeds in place. On such slopes a heavy rain would wash all the seeds downslope. If the slope has stones or ridges of some sort, the seeds are much less likely to move downslope.

Species Selection

The tree species chosen for direct seeding should be one that has been used successfully in the mine locality. If several different species have proved successful in the locality, then further selection criteria can be used. It is usually wise to choose the species that has the greatest economic importance for the region. It is also imperative that the species be adapted to the soil to be planted. On several occasions southern pine seeds have been sown on alkaline minesoils. Any seedlings that established on these minesoils usually showed poor growth and the light yellow coloration of iron deficiency.

Seed Quality, Storage, and Treatment

Certain factors should be known about the seeds purchased for direct seeding. These factors are percent viability, impurities present, percent of sound seeds, number of seeds per unit weight, and moisture content of seeds. Viability is the capacity of a seed to germinate. Sound seeds are those that have live tissue inside the seed coat. A seed still may not germinate due to seed dormancy, even though it is sound. Impurities are any foreign materials found in the purchased seeds. Impurities can be stones, leaves, soil, stems, or any other foreign matter. Moisture content of the seeds is the percent of the seed weight composed of water.

Seed viability should be known in order to properly determine the amount of seed to sow to obtain the desired number of trees per acre. If seed viability is known and the number of seed per pound is known, the number of possible seedlings resulting from the direct seeding can be calculated. The actual number of seedlings resulting will depend on many factors and experience is required to obtain this figure. Another factor that can affect the actual number of seedlings is the amount of impurities present. Impurities should be kept to a minimum in order to obtain the desired numbers of seeds per pound when weighing seeds for sowing.

If seeds are not sown immediately they should be stored carefully and promptly until used. This usually means that the seeds should be refrigerated at a temperature just above freezing. Most seeds should be stored in a relatively dry condition, but a few seeds require moist conditions. The seeds of many tree species quickly lose viability at high temperatures and moist conditions. The best moisture content for storing most tree seed is between 5 and 12 percent. However, acorns should be kept at a moisture content above 30 percent and beech seeds at a moisture content of about 17 percent. Seeds that have been treated by moist stratification must also be stored in a moist condition at a temperature slightly above freezing until sowed. Seeds can be killed by reducing their moisture content below some critical level, but this level is not known for most species.

When sound seeds do not germinate after being placed in the proper conditions of moisture, light, and temperature, they are said to be dormant. This dormancy can be caused by some physical characteristic such as a thick seed coat that does not allow moisture to enter the seeds or it may be a physiologic characteristic of the seeds. Such seeds can usually be induced to germinate by duplicating the conditions that nature uses to break dormancy. In the case of a thick seed coat, the coat may be thinned by abrasion or acid treatment. Physiologic dormancies are often broken by subjecting the seeds to low temperatures under moist conditions for a period of time or by treating the seeds with chemicals.

Most tree seeds are eaten by birds, rodents, and insects. They are also attacked by disease organisms. To alleviate these problems, many seeds are coated with chemicals before sowing, and others are covered with a layer of soil. The most successful seed coating for southern pine has probably been a mixture of Arasan (Thiram), Endrin, and latex. Arasan acts as a repellent for birds and Endrin repels rodents, other than squirrels. These repellents have also proved effective in other parts of the country. Repellents have not worked well on some smaller predators, such as some insects. Additional seeds are sown to compensate for these losses.

Seed Sowing

There are many methods of sowing, from hand sowing to helicopter sowing (Figure 20). Each method has some advantages when used under prescribed circumstances. Fixed-wing aircraft and helicopters can be used to sow large acreages quickly and economically. Hand-operated cyclone seeders have been used on large and small areas with success and economy. With certain seeds and under certain conditions, it may be necessary to hand plant seeds on small plots that have been scalped of competing vegetation.

It seems likely that one of the best methods of sowing tree seeds on minesoils will be with some type of seed drill drawn by a farm tractor. This method would allow for seeding in rows and would probably eliminate the need for repellents, as the seeds could be covered with a thin layer of soil. This thin layer of soil would also act as a mulch to keep the seeds more moist than they would be if lying on the soil surface. Finally, seed drilling would require fewer seeds per unit area than broadcast seeding.

Figure 20. Helicopter sowing tree seeds.

There are three basic positions in which the seeds rest after planting. Broadcast seeding leaves the seeds resting on the soil surface. This means that the seeds are exposed to predators. It also means that the seeds are less likely to obtain enough moisture for germination and seedling establishment because the soil surface is the first layer of soil to become dry. Seed broadcasting can be done by hand casting, hand-cyclone seeder, tractor-drawn cyclone seeder, fixed-wing aircraft, or helicopter.

In some sowing operations the seeds are pressed into the surface of the soil after sowing. This gives the seeds a greater chance to escape predators and to obtain needed moisture. Seeds that are pressed into the surface soil and lay on the soil surface would probably benefit from a predator-repellent coating. There are several planting implements that place the seeds on the soil surface and then push the seeds into the soil surface with a packing wheel. Seeds can also be dropped onto the soil surface by hand and then pushed into the soil surface by foot.

The final seed positioning method covers the seeds to a specific depth with a layer of soil. This method is more likely to keep the seeds moist for a longer period than the other two methods. It also gives the seeds more protection from predators. The two problems that occur with this method are placing the seeds too deeply and the possibility of a crust forming at the soil surface. If the seeds are planted too deeply the seedling may not push through to the soil surface, and if a crust forms they may not be able to break through. Seeds can be planted with a covering of soil by using various hand-operated and tractor-drawn implements.

Advantages and Disadvantages of Direct Seeding

There are several advantages and disadvantages of direct seeding when compared with seedling planting. Some of the following are interrelated, but they are listed individually to highlight the specific advantage or disadvantage.

Advantages of direct seeding:

1. Lower initial cost
2. Faster and can cover more acreage when the planting season is short
3. Less labor required
4. Less heavy equipment required
5. A natural root system develops whereas a transplanted seedling may have a deformed root system
6. Less site preparation may be required
7. Suitable for inaccessible areas that cannot be planted with seedlings
8. Often more successful for some tree species such as longleaf pine

Disadvantages of direct seeding:

1. Less control over stocking
2. Less control over spacing
3. Direct seeded stands may need precommercial thinning
4. Seeds may wash away on steep slopes
5. Newly germinated seedlings may be less resistant to competition than transplanted seedlings
6. Seeds and newly germinated seedlings are more susceptible to predators and drought

SEEDLING PLANTING

The most important effort to be made in a seedling planting program is to be certain that all phases of the operation are performed correctly. There is no need to plant seedlings correctly if the seedlings have been allowed to dry before reaching the planting site. There is no need to keep seedlings cool if they are to be planted on unsuitable soil. In other words, if the reclamationist does not perform all phases of the planting operation with care, the whole operation is likely to result in failure.

Seedling Storage

Seedlings should be obtained from a reliable nursery of good reputation. If the seedlings reach the reclamationist in a damaged condition, the entire planting operation is wasted. Seedlings should be shipped under conditions such that they are not allowed to heat, dry, or freeze. Even though the seedlings shipped to the planting site are dormant, the seedling packages produce small amounts of heat; if this heat is not allowed to escape, it will accumulate to a point at which the seedlings are killed. Because of this, seedling bales should have air circulating freely about them except for short periods of time. Seedlings are usually shipped in a moist medium that should not be allowed to dry. Freezing is also a problem during some cold periods. If small roots are frozen, they are killed and do not function to provide water and nutrients for the seedling when planted.

Seedlings should be planted as soon as possible after they leave the nursery. However, they often must be stored at the planting site due to delays or other problems. It is best to store the seedlings in their original containers if they can be kept cool, aerated, and moist. Many large seedling planting operations now use refrigerated rooms for seedling storage. Such storage will keep the seedlings in good condition for a reasonable period at a temperature slightly above freezing. If cold storage is not available, seedling bales or packages can be stored in any cool area up to about 4 weeks. The bales should be examined frequently to be sure that the seedling roots remain moist, aerated, and cool. Care must be taken to avoid problems resulting from lack of aeration when seedlings are stored in boxes. If the bales become too dry, they can be watered by several methods. However, be sure that the three water drains from the bale and air can reach the moist roots. Seedling roots submerged in water for as little as 24 hours may die.

If there are no facilities for storing packaged seedlings, the seedlings may be removed from the packages and heeled-in. Heeling-in is a storage method in which a trench is dug in moist, but not saturated, soil to a depth that will allow the roots to stretch to their full length. The trench should be dug in a place that will be shaded from the sun and protected from wind. Seedlings are placed in the trench and covered with soil up to the root collar. The soil around the seedlings should be kept moist but not saturated. Under good conditions, heeled-in seedlings will remain healthy for 1 to 2 months.

Hand Planting

There are two basic methods of planting seedlings on minesoils. The first and oldest method is hand planting. There are several tools and methods for hand planting, but the most probable ones for surface mine reclamation are the planting bar (dibble) and mattock. Faster planting can probably be done with a planting bar than any other hand tool and the results are good when the planting is done correctly. Approximately 800 to 1500 seedlings per worker-day can be planted with a planting bar. Types of mattock are preferred by some planters and are probably better for use on steep and moderately steep slopes.

As usual, soil moisture plays an important part in the success of hand planting. The planting bar or mattock is used to open a hole in the soil. The seedling roots are placed in the hole and arranged so that they fall straight down without turning upward or to the side. The hole is then closed completely without leaving any air pockets around the roots. All roots need to be in contact with moist soil in order to absorb enough water and nutrients to keep the seedling alive until the new roots are formed. Seedlings should be planted to the same depth they grew in the nursery or a little deeper. It should be remembered that the deeper the roots penetrate, the more the soil has to dry before the seedling dies of drought.

Some tree species are planted as cuttings rather than seedlings. Two of these species that have been planted on minesoils are cottonwood (*Populus deltoides*) and hybrid poplar (*Populus* spp.). Cuttings can be planted with any tool that will make a hole in the soil, such as a planting bar. However, the most popular tool for planting cuttings is a round iron bar with a footrest for pushing the bar into the soil. The bar should be 0.5 to 0.75 inch in diameter and should make a hole about 14 inches deep. A minimum length of 10 to 12 inches is recommended for cuttings. All of the cutting is placed in the planting hole except for the topmost bud. Again the soil must be in close contact with that portion of the cutting that is placed in the soil. It is also imperative that cuttings be kept moist from the time they are cut until they are planted.

Many minesoils are too compacted or stony for successful hand planting. Under these circumstances it has been found helpful to plow lines with a subsoiler or ripper and plant the seedlings or cuttings in these lines. These lines probably provide better water relations for the seedlings and better root development. The disadvantages are the expense and the likelihood of pulling large stones to the soil surface.

Machine Planting

The second basic tree planting method is machine planting. Many mechanical tree planters have been developed since the 1940s for many different planting situations. The simplest and most lightly constructed machines were developed for use on old fields where there is little vegetation or stone to interfere with the planting operation. Heavier and more complicated machines have been developed for planting in rough, vegetated, or stony soils.

Most of the tree-planting situations on minesoils should be easy ones under the present federal reclamation regulations. There should be no stones or woody plants on the minesoil. Therefore, the simpler and lighter tree planters should be adequate if the minesoil is not too compacted. Mechanical tree planters plant from 5000 to 15,000 seedlings per day, depending on planting conditions. This is not true when reclaiming older minesoils that were created before the replacement of topsoil was required. These older minesoils probably contain many stones and prevent the operation of a light tree planter. Some heavier planters will operate adequately if there are not too many large stones. There are a few planters designed to plant seedlings on stony soils. These planters operate on a principle similar to a planting bar. If the mechanical bar strikes a stone or other hard object, the bar retracts, recycles, and keeps pushing into the soil until a spot is reached where the seedling can be planted. These machines are expensive and their maintenance is probably more expensive than traditional tree planters.

Traditional tree planters, with coulter and plowshare, can be fitted with plows to scalp any competing sod from the area where the seedling is to be planted. Care must be taken to avoid soil erosion when sod is scalped. Various plows and devices have been fitted onto the tractor itself to clear debris, smooth the soil surface, and scalp competing sods. These competing sods can also be killed with herbicides to reduce the competition that they offer to newly planted tree seedlings. Sod roots will be left in place when sods are killed with herbicides. These roots will prevent soil erosion for a significant period.

Seedling Planting Precautions

There are several precautions to be taken with any seedling planting operation:

1. Be sure that the seedlings are healthy and of a proper size before planting them. It is always best to reject unsuitable seedlings rather than waste time, money, and reputation on a planting job that does not have a reasonable chance for success.
2. Keep the roots moist and cool, but not frozen, until the seedling is planted.
3. Roots must often be pruned to obtain proper root placement, but be sure that the planting personnel are not needlessly cutting roots or twisting them together in order to ease the planting job.
4. Plant the seedlings as deeply as allowed by the root system or planting tool. The more root in contact with moist soil, the greater chance for seedling survival. Planting holes or slits should be at least 8 inches deep.
5. Be sure that the seedling roots are straight in the planting hole and not turned upward or sideways.
6. Close the planting hole or slit completely. Do not leave air pockets around the roots. All roots must be in firm contact with moist soil in order to improve chances of seedling survival.

7. Seedling spacing must be checked constantly. There is a strong tendency for hand or machine planters to plant seedlings too closely. Close spacing can result in a waste of seedlings and a greater amount of time for trees to reach commercial sizes. Seedlings planted too far apart may result in too few seedlings per acre to satisfy reclamation regulations and a loss of economic return from wood production.

REFERENCES

U.S. Department of Agriculture, Forest Service. 1961. *Handbook on Soils,* Chapt. 6, p. 148. Washington, D.C.: Government Printing Office.

U.S. Department of Interior and U.S. Department of Agriculture. 1959. *Irrigation on Western Farms,* Agr. Inf. Bul. No. 199. Washington, D.C.: Government Printing Office.

SUGGESTED READING: PART I

Buckman, H.O., and N.C. Brady. 1960. *The Nature and Properties of Soils.* 6th ed. New York: Macmillan.

Donahue, R.L., J.C. Shickluna, and L.S. Robertson. 1971. *Soils: An Introduction to Soils and Plant Growth.* Englewood Cliffs, NJ: Prentice-Hall.

Jones, U.S. 1979. *Fertilizers and Soil Fertility.* Reston, VA: Reston Publishing.

Mengel, K., and E.A. Kirkley. 1982. *Principles of Plant Nutrition.* 3rd ed. Bern: International Potash Institute.

Russell, E.W. 1973. *Soil Conditions and Plant Growth.* 10th ed. London: Longman.

Tisdale, S.L., and W.L. Nelson. 1975. *Soil Fertility and Fertilizers.* 3rd ed. New York: Macmillan.

Thomson, J.R. 1979. *An Introduction to Seed Technology.* New York: Wiley.

PART II

THE PRACTICE OF REVEGETATION

INTRODUCTION

The revegetation of any drastically disturbed land, such as a surface mine, is probably the single best method of stabilizing the soil. Vegetation is cheaper than any other form of soil stabilization because of the continuous and long-term effect. It is also more efficient than any other method and more aesthetically pleasing. Finally revegetation can place the soil into continued economic production.

Vegetation prevents or reduces soil erosion by providing a soil cover that intercepts raindrops and prevents them from dislodging soil particles and destroying soil structure. The plant roots bind soil particles together and prevent water from carrying soil downhill. Above-ground vegetation slows water runoff along the soil surface and enables more of the water to move into the soil for plant use. When this vegetation intercepts water flowing over the soil surface it also reduces the downhill movement of soil.

This discussion of surface mine revegetation assumes that the area to be vegetated has been regraded and all land leveling completed. This is true for overburdens and minesoils. The author has attempted to take the revegetation steps in their normal sequence and discuss each of them in the order in which the reclamationist will apply them.

CHAPTER 7

SOIL SAMPLING, ANALYSIS, AND RECOMMENDATIONS

Soil testing or analysis has been used for many years to determine the kinds and amounts of nutrients in a soil. Laboratories can provide soil analyses for acidity, alkalinity, soluble salts, all the essential elements for plant growth, toxic elements, cation exchange capacity, organic matter content, soil-water relations, and practically any other analysis that might be needed. However, these analyses are only as good as the samples sent to the laboratory and the competence of the laboratory. Therefore, it is wise for the reclamationist to collect valid samples and send them to a reliable soil testing laboratory.

The greatest problem with soil testing and minesoils is the fact that the analytic methods, correlations between analyses and plant growth, and fertilizer recommendations are all based on natural soils. Some minesoil properties are different from natural soils and require new and special laboratory techniques. These techniques are being developed and will assume their place in the near future. In the meantime, present techniques give valuable information and should be used whenever possible.

SOIL SAMPLING

Soil samples should be taken from the minesoil and sent to the testing laboratory in time to receive the test results before purchasing limestone and fertilizer. Soil testing laboratories have busy and slack seasons. Therefore, the laboratory should be asked the amount of time they will require for soil testing in order to allow sufficient time for fertilizer purchases and application prior to the planting season.

It is necessary to divide the area to be planted into sections that appear uniform in soil color and texture (Figure 21). Soils that vary in texture or color are likely to vary in nutrient content. Dividing areas into uniform sections for sampling may be easier on surface mines than on natural soils because minesoils are humanmade and the uniformity of

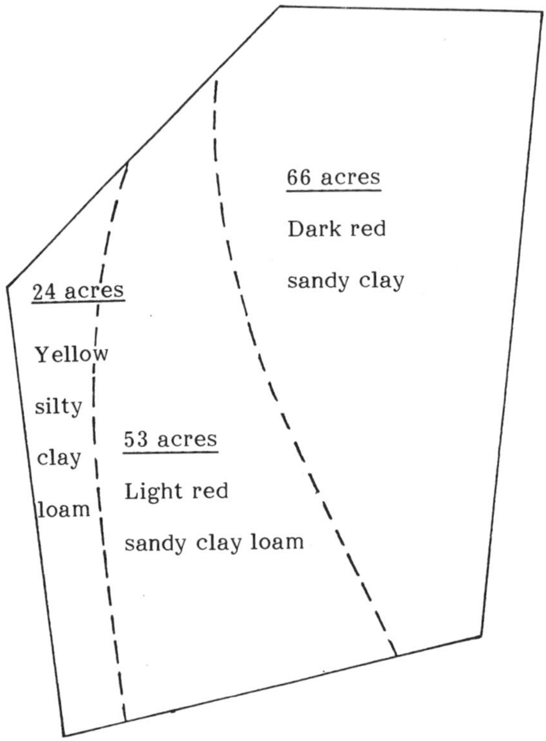

Figure 21. This topsoiled area is divided into sections according to minesoil color and texture.

the replaced topsoil is partially known. However, some of these uniform sections may be quite large and it would probably be wise to limit the size of any section to be sampled to approximately 25 acres (Figure 22). This limit prevents the application of erroneous test results to large acreages.

Approximately 20 subsamples are taken from each section (Figure 22). These subsamples can be taken in a random fashion that covers the whole section, or the section can be divided into 20 approximately equal parts and a subsample taken from each part. The random method is illustrated in Figure 22.

Each subsample should be taken to a uniform depth of between 6 and 9 inches, with care used to be sure that all subsamples are of the same approximate volume. After placing the subsamples in a clean container, they are mixed thoroughly. The sample to be sent to the laboratory is taken from this mixture. This sample can be any size requested by the laboratory but is usually about a pint. Each sample is identified and a record kept of where each was taken in order to judge the effectiveness of soil amendment programs.

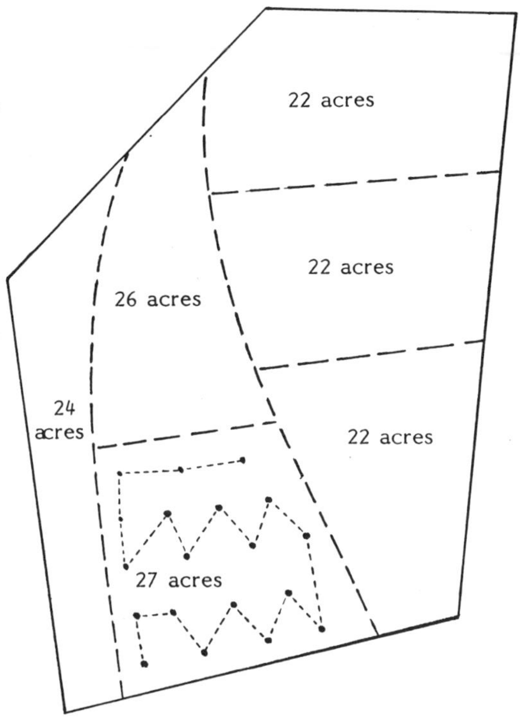

Figure 22. Sampling units from Figure 21 are further subdivided into approximately 25-acre sampling units. Possible locations of soil subsamples are represented by dots.

It is probably best to find a reliable soil testing laboratory and use it exclusively. Methods and results often differ among laboratories and it is easier for the reclamationist to develop a good fertilization program if the test information received over a period of years is obtained by similar analytic methods.

SOIL ANALYSIS

Soil testing laboratories are usually equipped to perform many different soil analyses. It is unlikely that the reclamationist will need all the available analyses. Analyses have two major purposes: to tell the reclamationist the kinds and amounts of fertilizer elements needed to grow a specific crop on a specific soil; and to determine whether the soil is acid, alkaline, or neutral and the amount of limestone needed to correct any soil acidity. The soil nutrient analyses most often needed by the reclamationist are for soil potassium and phosphorous contents that will be available for the desired plants. Of course, nitrogen is also important to plant growth, but soils are usually deficient in this element and it must be added to

Table 10. Ranges of pH for Best Growth of Selected Plants

Plant	pH	Plant	pH
Alfalfa	6.5–7.5	Grasses	5.5–7.0
Sweet clover	6.5–7.5	Oats	5.5–7.0
Barley	6.5–7.0	Sudan grass	5.5–7.0
Birdsfoot trefoil	6.2–6.5	Bermuda grass	5.5–7.0
Alsike clover	6.0–7.0	Kentucky bluegrass	5.5–7.0
Red clover	6.0–7.0	Sorghum-sudan	5.5–6.5
Peas	6.0–7.0	Tall fescue	5.5–6.5
White clover	6.0–7.0	Millet	5.5–6.5
Crownvetch	6.0–7.0	Lespedeza	5.5–6.5
Timothy	6.0–6.5	Rye	5.5–7.0
Johnson grass	6.0–8.0	Sorghum	5.5–7.0
Wheat	5.5–7.0	Soybeans	5.5–7.0
Corn	5.5–7.0	Vetch	5.5–6.5
Cowpeas	5.5–7.0	Buckwheat	5.2–6.5
Crimson clover	5.5–7.0	Redtop	4.5–6.0

practically all soils to obtain acceptable plant growth. Therefore, the testing laboratory will usually recommend the addition of standard amounts of nitrogen, depending on the crop to be grown. The other analysis that is vitally needed is the soil pH. This analysis tells whether the soil is acid, neutral or alkaline. If the soil is alkaline (pH > 7.5), neutral (pH 6.5 to 7.5), or mildly acid (pH 6.5 to 5.5), the reclamationist will probably not change the pH but will use this information to determine which crop will be grown and how it will be fertilized. If the pH shows that the soil is moderately (pH 5.5 to 5.0) or severely acid (pH < 5.0), a test for lime requirement can be made. This test will tell the approximate amount of finely ground limestone needed to bring the soil pH up to some approximate pH such as 6.5. Raising a soil pH from 5.0 to 6.5 can significantly increase the availability to plants of such essential elements as nitrogen, phosphorous, potassium, sulfur, calcium, and magnesium (see Chapter 1, Figure 10).

Among the first items of information needed to begin a revegetation project are the pH, amount of lime needed, and amounts of nutrients needed for the minesoil that is to be vegetated. The pH is needed to decide which plants should be grown. Most plants perform best within a definite soil pH range (Table 10). This does not mean that these plants will not survive and grow in other pH ranges. However, it is not reasonable to try to establish plants that may not do well at a given pH when other plants have a better chance of establishing and growing satisfactorily. After all, one of the major objectives of reclamation is to completely cover the soil surface with vegetation as quickly as possible in order to control erosion.

Table 11. Approximate Amounts of Finely Ground Limestone Needed to Raise Top 6 to 8 Inches of Minesoil to pH 6.5

	Limestone needed (ton/A)					
	Original pH 3.5		Original pH 4.5		Original pH 5.5	
Soil texture	**Cool[1]**	**Warm[2]**	**Cool**	**Warm**	**Cool**	**Warm**
Sand	1.5	1.0	1.1	0.7	0.6	0.4
Loamy sand	1.5	1.0	1.1	0.7	0.6	0.4
Sandy loam	2.8	1.7	2.1	1.2	1.3	0.7
Loam	4.0	2.6	2.9	1.8	1.7	1.0
Silt loam	4.9	3.8	3.5	2.6	2.0	1.4
Clay loam	6.0	5.0	4.2	3.5	2.3	2.0

[1]Cool region, see Figure 23, region 2.
[2]Warm region, see Figure 23, region 1.
Source: Based on U.S.D.A. Soil Conservation Service. U.S.D.A. Handbook No. 18. 1951, p. 237.

Figure 23. Soil regions for limestone additions described in Table 11.

There is no economical method of lowering a soil pH that is considered too high for the growth of a particular plant. Plants that will perform well at a high soil pH are selected for growth on these soils of high pH. However, if the soil pH is lower than desired for a particular plant, it can be raised by incorporating finely ground limestone or some other alkaline material. The amount of limestone needed to raise the pH a given amount is usually different for different soils. Soil testing laboratories can determine amounts of limestone needed for this purpose within acceptable limits. Table 11 shows the approximate amounts of limestone needed to raise the pH of soils of different textures. The limestone needs determined by soil testing laboratories are more likely to be accurate than the values shown in this table. This table should only be used when laboratory results are unavailable.

If the minesoil is too acid for good plant growth, a specific amount of agricultural grade limestone will be recommended by the soil testing laboratory to counteract the acidity. The limestone must be ground finely enough so that at least 50 percent will pass through a 60-mesh sieve and 90 percent will pass through a 10-mesh sieve. The amount of lime needed to raise soil pH to an acceptable level is determined by adding a buffer solution to a soil sample and then measuring the soil pH. This method works well for normal agriculture soils or for the topsoils placed over overburdens. However, overburden acidities may be caused by other factors and may need to be determined by other methods. These high overburden acidities are usually caused by iron sulfide (FeS), which is also called iron pyrite or fool's gold. The sulfur in this mineral oxidizes to become sulfate (SO_4), which combines with water to form sulfuric acid (H_2SO_4). The amount of sulfur in the iron sulfide can be determined by chemical methods and the amount of sulfuric acid that could be formed from the sulfur is then calculated. When the amount of sulfuric acid that could be formed is known, the amount of limestone needed to neutralize this much acid can be calculated. However, it is seldom worthwhile to add more than enough limestone to correct present acidity. More limestone can be added later as more pyrite oxidizes.

The addition of limestone to an acid soil is often omitted by reclamationists. This is a serious mistake because there is usually little value in adding fertilizer without correcting soil acidity. Also, toxic elements in acid soils must often be neutralized by liming before plants can establish or survive. Failure to lime an acid soil will often result in a vegetation failure.

RECOMMENDATIONS

The soil testing laboratory recommendations will prescribe limestone in the tons per acre needed to counteract present acidity. This amount of limestone should be spread uniformly over the minesoil surface and incorporated with the minesoil to a depth of at least 6 inches. If a topsoil or overburden is extremely acid, a greater volume of soil will become available for root growth if lime is mixed to a greater depth such as 12 inches. Of course, this would require twice as much lime per acre. Limestone can be incorporated into the minesoil by discing, spring-tooth harrowing, or any other convenient method such as using ripper teeth attached to a bulldozer. The more evenly and thoroughly the limestone is incorporated throughout the soil profile, the more quickly and completely the soil acidity will be reduced.

It would probably be best to apply and mix the limestone and soil from 1 to 6 months before planting. However, this is often difficult for the reclamationist. If the limestone and soil are well-mixed by plowing, discing, or harrowing, seeding can take place immediately. If mixing is not thorough, better results would be obtained by waiting the 1- to 6-month period.

The two most common materials used to reduce soil acidity are calcitic limestone and dolomitic limestone. Dolomitic limestone provides both calcium and magnesium to the soil. However, it is believed that the addition of magnesium to some minesoils, already high in magnesium, is harmful. Calcitic limestone will neutralize soil acidity as well as dolomitic, and provides calcium but not magnesium to the soil. The soil testing laboratory should be able to recommend the preferred type of limestone.

Soil testing laboratories will recommend the nutrients needed for the desired crop. Their recommendations are based on the mineralogy and nutrient status of the soil in question and the crop to be grown on this soil. In most cases they will recommend that nitrogen, phosphorous, and potassium be applied. These are the most common nutrients needed for grasses and legumes. However, there are some cases where other nutrients are needed such as zinc, copper, manganese, iron, sulfur, molybdenum, and boron. A special request must be made to the laboratory in order to obtain these analyses if such a deficiency is suspected or if such a deficiency commonly occurs in the area.

Nitrogen is probably the most troublesome nutrient because it remains in the soil for only short periods of time. It is therefore best to assume that any nitrogen fertilizer is depleted within 45 days after application. Soil nitrogen is lost through plant uptake, rainwater leaching, use by microorganisms, and volatilization. Nitrogen is stored in organic matter but new minesoils have little organic matter. To compensate for the rapid loss of applied nitrogen, it is helpful to apply half of the nitrogen fertilizer before seeding and the other half 30 or 45 days after seeding. Splitting the applications in this fashion makes more of the nitrogen fertilizer available to the plants. The type of nitrogen fertilizer applied is usually not important. The most likely types to be encountered are ammonium nitrate (NH_4NO_3), aqueous ammonia (NH_4OH), ammonia gas (NH_3), and urea ($(NH_2)_2CO$). The price per unit of nitrogen is the factor of most concern to the reclamationist because most nitrogen fertilizers are about equal in fertilizing values. The price paid per pound of nitrogen can be calculated by knowing the percent of nitrogen contained in the fertilizing material.

Table 12. Amounts of Nitrogen Fertilizers Needed (lb/A) to Satisfy Soil Testing Laboratory Recommendations for Nitrogen (N)

N recommendation	Ammonium nitrate (33)[1]	Urea (45)	Anhydrous ammonia (82)	Nitrogen solution (30)	Ammonium sulfate (21)
10	30	22	12	33	48
20	61	44	24	67	95
30	91	67	37	100	143
40	121	89	49	133	190
50	151	111	61	167	238
60	182	133	73	200	286
70	212	155	85	233	333
80	242	178	98	267	381
90	273	200	110	300	428
100	302	222	122	333	476
110	333	244	134	367	524
120	364	267	146	400	571
130	394	289	158	433	619
140	424	311	171	467	667
150	454	333	183	500	714
160	485	355	195	533	762
170	515	378	207	567	809
180	545	400	220	600	857
190	576	422	232	633	905
200	606	444	244	867	952

[1]Numbers in parentheses indicate percent nitrogen.

Phosphorous is the next most likely nutrient to be needed. All phosphorous fertilizers contain phosphorous in a readily available form except rock phosphate. Phosphorous is not lost from the soil as readily as nitrogen, but it does convert into unavailable forms (see Chapter 1, Figure 10). The most likely forms of phosphorous fertilizer to be encountered by the reclamationist are superphosphate (20 percent P_2O_5), triple superphosphate (45 percent P_2O_5), concentrated superphosphate (50 percent P_2O_5), diammonium phosphate (46 percent P_2O_5), and monoammonium phosphate (48 percent P_2O_5). Superphosphate, triple superphosphate, and concentrated superphosphate all contain calcium and sulfur. The ammonium phosphates do not contain sulfur or calcium, but they are a good source of nitrogen in addition to phosphorous. The phosphorous content of a soil can be increased to dangerous levels over a period through excessive fertilizer applications. This is true because the excess phosphorous may combine with some other essential element and drastically reduce the availability of the other essential element. Soil tests should be made periodically to detect such a phosphorous buildup.

Table 13. Amounts of Phosphorous Fertilizers Needed (lb/A) to Satisfy Soil Testing Laboratory Recommendations for Phosphorous (P_2O_5)

P_2O_5 recommendation	Ordinary or super-phosphate (20)[1]	Concentrated or triple super-phosphate (48)	Mono-ammonium phosphate (48)	Diammonium phosphate (46)
10	50	21	21	22
20	100	42	42	43
30	150	62	62	65
40	200	83	83	87
50	250	104	104	109
60	300	125	125	130
70	350	146	146	152
80	400	167	167	174
90	450	187	187	196
100	500	208	208	217
110	550	229	229	239
120	600	250	250	261
130	650	271	271	283
140	700	292	292	304
150	750	312	312	326
160	800	333	333	348
170	850	354	354	369
180	900	375	375	391
190	950	396	396	413
200	1000	417	417	435

[1]Numbers in parentheses indicate percent phosphorus.

Three arbitrary forms have been assigned to the potassium found in soils. Most soil potassium is usually in a form that is unavailable to plants; some is in a form that slowly becomes available to plants; and some is in readily available form. The form that is applied in fertilizers is readily available. However, under some circumstances it will change into the less available forms over a period. Some of the readily available potassium is taken up by the crop and some may be lost by rainwater leaching. The most likely form of potassium fertilizer to be purchased is potassium chloride (KCl) (60 percent K_2O).

All the above nutrients (nitrogen, phosphorous, and potassium) can be purchased separately or blended in practically any combination. Blended fertilizers are labeled with the percent of each nutrient in the blend. Somewhere on the label will be a series of three numbers, such as 10-10-10, 0-20-10, or 8-15-20. The first number on the label is the percent of nitrogen (N) in the blend. This is sometimes called elemental

Table 14. Amounts of Potassium Fertilizers Needed (lb/A) to Satisfy Soil Testing Laboratory Recommendations for Potassium (K_2O)

K_2O recommendation	Muriate of potash or potassium chloride (KCl) (60)[1]	Potassium sulfate (K_2SO_4) (50)	Potassium magnesium sulfate ($K_2SO_4 \cdot MgSO_4$) (22)	Potassium nitrate (KNO_3) (44)	Potassium carbonate (K_2CO_3) (57)
10	17	20	45	23	17
20	33	40	91	45	35
30	50	60	136	68	53
40	67	80	182	91	70
50	83	100	227	114	88
60	100	120	273	136	105
70	117	140	318	159	123
80	133	160	364	182	140
90	150	180	409	204	158
100	167	200	454	227	175
110	183	220	500	250	193
120	200	240	545	273	210
130	217	260	591	295	228
140	233	280	636	318	246
150	250	300	682	341	263
160	267	320	727	363	281
170	283	340	773	386	298
180	300	360	818	409	316
190	317	380	864	432	333
200	333	400	909	454	351

[1]Numbers in parentheses indicate percent potassium.

nitrogen. The second number is the percent of P_2O_5, which represents the phosphorous. The P_2O_5 is sometimes called available phosphoric acid. The third number is the percent of K_2O, which represents the potassium. The K_2O is often called potash.

If you wish to convert the K value to K_2O, multiply K by 1.20. To convert K_2O to K, multiply K_2O by 0.83. To convert P to P_2O_5, multiply P by 2.29. To convert P_2O_5 to P, multiply P_2O_5 by 0.44.

Some of the blends encountered in different parts of the U.S. are: 8-8-8, 10-10-10, 13-13-13, 4-8-16, 15-0-15, 4-12-12, 5-10-10, 0-20-20, and 10-9-0. Normal superphosphate is 20 percent P_2O_5 but has no nitrogen (N) or potash (K_2O) and would be labeled 0-20-0. Ammonium nitrate (NH_4NO_3) is labeled 33-0-0, and muriate of potash is labeled 0-0-60.

If you buy 100 pounds of a blend labeled 8-14-20, it will contain 42 (the sum of 8-14-20) pounds of fertilizer and 58 pounds of other chemical elements, sand, limestone, or some other filler. Under these circumstances, it is worthwhile for the reclamationist to give some thought to the fertilizers being purchased.

Fertilizers can be bought unblended and the cost of fillers plus the transportation of fillers avoided. For instance, nitrogen can be bought as ammonium nitrate (NH_4NO_3), which is 33 percent nitrogen (N). Phosphorous can be bought as normal superphosphate, which is 20 percent P_2O_5; and potassium as muriate of potash (KCl), which is 60 percent K_2O.

If 200 lb/A of 20-10-30 are needed to fertilize a particular soil, the amounts of ammonium nitrate, normal superphosphate, and muriate of potash needed to supply these nutrients can be calculated as follows:

20 percent of 200 pounds = (0.20 x 200) = 40 pounds of N needed

10 percent of 200 pounds = (0.10 x 200) = 20 pounds of P_2O_5 needed

30 percent of 200 pounds = (0.30 x 200) = 60 pounds of K_2O needed

Amount of N in NH_4NO_3 = 0.33

Therefore: 40 ÷ 0.33 = 121.1 pounds of NH_4NO_3 needed per acre

Amount of P_2O_5 in normal superphosphate = 0.20

Therefore: 20 ÷ 0.20 = 100.0 pounds of normal superphosphate needed per acre

Amount of K_2O in muriate of potash = 0.60

Therefore: 60 ÷ 0.60 = 100 pounds of muriate of potash needed per acre

See Tables 12, 13, and 14 for the amounts of fertilizer to apply to the soil in order to satisfy the nitrogen, phosphorous, and potassium recommendations of the soil testing laboratory.

CHAPTER 8

SEEDBED PREPARATION

The preparation of a proper seedbed is one of the most important factors in successful land reclamation. Methods of preparing a seedbed vary with the type of seeds, soil, climate, time of planting, and available equipment. Certain principles must be understood in order to cope with these variations and produce an acceptable stand of vegetation. A seedbed should be prepared in such a way that the soil will provide sufficient moisture, acceptable temperatures, optimum fertility, and proper tilth for successful seed germination and plant establishment. Much of this success depends on proper soil tillage.

SOIL TILLAGE

The purpose of soil tillage is to establish and promote a vegetative crop of some type. In land reclamation this crop is usually a forage crop or tree crop. Tillage operations should minimize soil erosion and promote soil moisture content, seed germination, and root growth. In row crops, tillage is also used to control weeds.

At one time it was believed that a maximum amount of tillage would produce the best crop. Today this is thought untrue and the trend is to do as little tillage as possible to meet objectives.

A seedbed should be loose enough for a new seedling root and air to penetrate easily but firm enough that the soil will have micropores to transmit water to germinate seed. Also close contact between the soil and seed is necessary for the soil to transmit water to the seed for the germination process. There are many tillage methods for preparing a seedbed and any method that forms a seedbed with the desired qualities is acceptable. Seedbed tillage can also be used to thoroughly mix soil amendments with the soil. Soil amendments are any material added to a soil, such as fertilizer, limestone, organic matter, sand, animal manure.

Figure 24. Moldboard plow.

PRIMARY TILLAGE

There are many types of plows and many purposes of tillage with these plows. In conventional farming operations the first or primary tillage operation is usually one that cuts and breaks the soil into relatively large clods. This operation penetrates the soil more deeply than succeeding tillage operations and it may also be used to incorporate soil amendments more deeply than succeeding tillage operations. The soil surface is usually left in a rough condition with many large open spaces between the lumps of soil. The primary tillage operation is done by moldboard or chisel plows, subsoilers or rippers, heavy disc plows, or rotary tillers. There are other implements used for this purpose, but these are the ones most frequently used. All of these implements, except the rotary tiller, leave the soil in a poor seedbed condition.

Moldboard plows are used to turn fine-textured (silty or clayey) soils or sods (Figure 24). They are designed for agricultural work rather than reclamation. These plows are pulled through the soil at a predetermined depth. They slice through the soil and curved wings (moldboards) turn the soil slice up and over. Moldboard plows do not function well on hard, sticky, or stony soils. They have relatively high power requirements and operate at slow speeds.

Figure 25. Chisel plow.

Chisel plows are used to stir and aerate the soil without actually turning the soil as a moldboard plow would do (Figure 25). Chisel plows do not require as much power as a moldboard plow unless the chisels are allowed to penetrate deeply into the soil. Chisel plows can be adjusted to barely scratch the soil surface or to dig as deeply as 16 inches. Also the chisels may be anything from a narrow point to wide sweeps that are used for weed control.

Chisel plows are suited to stony soils or soils with other obstructions. The narrow, flexible shanks that enter the soil bypass obstacles more easily than a moldboard plow. These chisel plows can prepare a reasonably good seedbed for broadcast seeding in some soils and they efficiently mix lime and fertilizer with the soil material. One disadvantage of these plows is that they have a tendency to pull stones from below the soil surface up to the surface. This could be a problem if they are allowed to plow deeply enough to pull stones from the overburden to the surface of the topsoil. An advantage of the chisel plow is that it does not compact soils as badly as disc plows and harrows.

Heavy disc plows are used for the same purpose as a moldboard plow but the disc has several advantages (Figure 26). The disc is more effective on heavy clay soils, hard soils, hardpans, and deep plowing. However, more power is required to pull the disc than the moldboard.

Disc plows have two or more discs aligned on a single axis or they may be mounted on individual axes. They can be adjusted hydraulically for different plowing depths, and on some models the pitch and angle of the discs can be changed. Disc plows can be used for deep plowing, aerating the soil, and mixing lime and fertilizer with the soil. They are

Figure 26. Disc plow.

more useful than moldboard plows for hard and sticky soils. They should not be used where there are many large stones. Disc plows usually require more power than chisel plows and operate at slower speeds. Discs compact soils more than moldboards and chisels.

Subsoilers and rippers are used to fracture compacted soils or hardpans that are below the normal tillage depth (Figure 27). Such soils usually require subsoiling in order to improve one or more of the following:

Figure 27. Subsoiler.

aeration, water infiltration, water storage, water drainage, and root penetration. Soil must be at the proper moisture content for subsoiling to be successful. If the soil is wet, subsoiling operations will probably fail because the wet soil will flow back into the cracks formed by the subsoiler. Subsoiling should be done when the soil is relatively dry in order to get a breaking or shattering action in the hard soil. If the soil is too dry, the soil may crumble and fill the formed cracks. However, some reclamationists prefer a very dry soil for subsoiling. Some experimentation may be required to determine the best moisture content for subsoiling a specific soil. Subsoilers have high power requirements.

The rotary tiller breaks ground and prepares the seedbed, all in one operation. This means that only one trip is made over the soil instead of the usual three trips for plowing, discing, and harrowing. However, the rotary tiller may require as much power for one operation as the total power required for three separate operations with conventional equipment. As primary tillage is not usually required on topsoiled minesoils, rotary tillage is likely to be more expensive than the usual secondary tillage alone. The rotary tiller may save time and this could be important when time and labor are limited. Rotary tillers should not be used on soils with large rocks or slopes of 30 percent or more.

SECONDARY TILLAGE

If a rotary tiller is not used, additional tilling must be done after primary tillage to reduce the large soil lumps, eliminate the large air spaces, level the soil surface, and make the soil firmer in order to produce a

Figure 28. Single-action disc harrow with centrifugal broadcast seeders mounted behind tractor cab.

Figure 29. Double-action disc harrow.

suitable seedbed. The implements used in this operation can be one or more of the following: disc harrow, spike-tooth harrow, spring-tooth harrow, tine harrow, roller harrow, and drag. Other implements are used to prepare the soil for seeding, but these are the ones used most frequently.

This second tillage operation should be done as quickly as possible after the first. The soil will tend to dry at the surface after the first operation and then will dry again after the second operation. If the second operation follows the first one closely, the total amount of soil moisture lost will be held to a minimum. It is also easier to break clods that are still moist from the first operation. Finally, the second operation will close many of the air pockets from which soil moisture can escape.

Disc harrows are probably the most versatile of all tillage implements. They can be used to incorporate soil amendments, control weeds, prepare seedbeds, and work lands where stones, roots, or stumps would inhibit other implements (Figure 28). Disc harrows are classified as single action, double action, or offset. All of these can be mounted either on a three-point hitch or pulled from a drawbar. A possible disadvantage of the disc is the soil compaction caused by the extreme pressure exerted by the thin edge of the disc on the soil. This compaction becomes serious if a disc is used repeatedly on the same soil.

Figure 30. Spike-tooth harrow.

Single-action discs are two sets of discs that operate side by side and can be set to throw the soil outward or inward from the center of the implement (Figure 28). For seedbed preparation they would be set to throw soil outward to either side of the tractor. Double-action discs consist of four sets of discs (Figure 29). The two front discs throw the soil outward and the two rear discs throw the soil back into its original position. This leaves a smoother seedbed than the single-action disc. The offset disc harrow has one set of discs moving the soil in one direction and a following set moving the soil in the opposite direction. This action causes the disc to move in an offset position behind the tractor. Large offset disc harrows can often be used instead of disc plows.

Lightweight disc harrows are not suitable for soils with large stones or soils with a rough surface. These discs will cut through minor vegetation but may be broken by woody vegetation or stones. However, heavy disc harrows are often used on stony and rough overburdens.

Spike-tooth harrows are used to break soil clods left by plowing, loosen soils crusted by rainfall, smooth seedbeds, and cover broadcast seeds (Figure 30). These harrows require little power and can be pulled behind other implements to eliminate an additional trip over the soil.

The angle at which the harrow teeth enter the soil can be adjusted. When the teeth are vertical (perpendicular to the soil) they will create the greatest soil disturbance. When the teeth are more nearly horizontal (parallel to the soil) they do not disturb the soil as much and can be used to cover seeds with a light cover of soil. However, spike-tooth harrows do not cover seeds to a uniform depth.

Figure 31. Spring-tooth harrow.

There is a variation of the spike-tooth harrow called a chain or flexible harrow. This harrow resembles a section of chainlink fence with teeth hanging from the fence surface. The tooth angle cannot be changed, but the harrow can be reversed and this in effect changes the tooth angle. This harrow can also be overturned and pulled with the teeth pointing upward. Used in this fashion the harrow becomes a float and can be used to smooth the soil surface or lightly cover seed. The flexibility of the chain harrow allows it to conform to the soil surface.

The spring-tooth harrow can dig more deeply into the soil than the spike-tooth harrow and operates from 3 to 6 inches in depth (Figure 31). The spring-tooth harrow is used to break soil clods that are not hardened and to loosen soil crusts. It will also close air pockets and level some soils enough to plant seed. This implement is intended for relatively stony soils and may prove to be useful on many minesoils. However, abundant large stones may cause breakage. Two general types of teeth are used on these harrows. The utility or seedbed tooth is used for seedbed preparation, and the weeding tooth is used to kill persistent weeds.

Tine-tooth harrows perform much the same functions as spring-tooth and spike-tooth harrows (Figure 32). However, the teeth are thinner and tend to vibrate as they are pulled through the soil. This vibration helps to break soil clods and helps to move the tines around soil obstructions such as stones. Tine-tooth harrows are often towed behind other implements to obtain a smoother seedbed.

Figure 32. A tine-tooth harrow at the rear of a chisel plow.

Roller harrows are also known as corrugated rollers, cultipackers, and soil pulverizers (Figure 33). The purposes of a roller harrow are to crush soil clods, firm the soil surface to improve moisture availability for seed germination, and level freshly plowed soils. These harrows consist of rollers at the front and rear of the implement with some type of harrow teeth between the front and rear sets of rollers. The harrow teeth may be spike, spring, or tine. The rollers may be serrated, sprocket, crowfoot, or solid. Roller harrows are usually heavier than tooth harrows and they compact the soil surface in preparation for a seedbed. A seeder can be attached at the rear or between the roller harrows in order to prepare a seedbed and sow in one operation.

The rollers may be used alone and are usually called cultipackers when used in this manner. They are sometimes used to roll over broadcast seed and push them into firm contact with the soil of the prepared seedbed. Some type of mulch can then be applied to the seedbed. This system places the seed in close contact with a firm seedbed and provides good moisture relations for seed germination. The mulch helps to hold moisture at the soil surface and prevent excessive soil surface drying.

Many harrows are equipped with a smoothing board, drag, or float at the rear of the implement to provide additional smoothing action. This attachment is a board attached to the rear of the harrow that drags across the surface of the soil. Drags can also be separate implements made from boards, slabs of iron, or old tires that are tied together in such a way that they will provide a smoothing action when dragged over the soil surface.

Figure 33. Roller harrows or cultipackers at front and rear of implement, with spring-tooth harrow between them.

The preceding discussion covered most of the tillage implements and types of tillage. It is obvious that no one reclamation project will need all of these implements. However, there is probably a place for any one of the implements on at least one reclamation project in the eastern U.S. The majority of the reclamation tillage is probably done with heavy disc harrows and chisel plows. Disc harrows are used to loosen compacted layers at the soil surface and prepare a seedbed for broadcast seeding. If deeper penetration is needed to shatter compacted layers beneath the soil surface, chisel plows can be used. Chisel plows are also better for tilling soils with large rocks, where chisels are less likely to be broken than the discs of a disc harrow. Both of these implements can be used to efficiently incorporate lime and fertilizer with soil.

If a seed drill is to be used instead of a broadcast seeder, a second tillage operation will probably be needed for the efficient operation of the seed drill. This is usually necessary to further reduce the soil clods and smooth the soil surface. The implement for this second tillage can be a light disc harrow, spike-tooth harrow, spring-tooth harrow, tine-tooth harrow, roller harrow, or any other implement that will provide a smooth soil surface for efficient seed drill operation.

CHAPTER 9

PLANTING

There are four basic types of planters used in agriculture. Row crop planters are used to plant seeds such as corn, sorghum, soybeans, and cotton in uniform rows with precise spacing between seeds within the rows. Transplanters are used to transplant cabbage, tomato, tobacco, and tree seedlings from a nursery to the field where they will grow to maturity. Seed drills and broadcast seeders are used to plant small grain, grass, legume, and some tree seeds. At present, surface mine reclamationists are interested mostly in broadcast seeders, seed drills, and tree seedling transplanters.

SEED GERMINATION FACTORS

Whatever the equipment or method used for planting, certain environmental factors must be taken into account in order to obtain satisfactory seed germination and seedling establishment. Seeds must have adequate supplies of moisture and oxygen and a definite temperature range, and some seeds must have light to germinate. Soils should have about 50 to 75 percent of their total field capacity of water for adequate seed germination. Some seeds will germinate when they absorb enough moisture to equal 25 percent of their total weight, whereas other seeds require 75 percent. Under these circumstances, it would be best for the reclamationist to plant seeds when adequate supplies of moisture are present or expected and to use cultural practices that conserve this moisture.

Practically all seeds require oxygen to germinate. This means that seeds planted too deeply may not receive enough oxygen to germinate. Also, too much water in the soil can exclude oxygen and prevent germination.

Soil temperature must be taken into account when planting seeds. Some seeds, such as wheat, oats, and rye, will germinate slowly at temperatures as low as 33°F. However, the optimum temperature for wheat germination is about 59°F. Thus for any specific seeds there is a range of temperatures at which adequate germination is obtained. In general, plants that grow in the cool season have seeds that germinate at lower temperatures than those that grow best in the warm season. Buckwheat, red clover, and perennial ryegrass are cool weather crops and their seeds will germinate well at temperatures as low as 45°F. Warm weather crops such as Bermuda grass and crotalaria require 85°F or greater for best germination. The best overall temperature for seed germination appears to be approximately 70°F. Practically all states have county agricultural agents or agricultural experiment stations that have information regarding the best time of year or soil temperature for planting various crops (Appendix A).

Most field crops are not sensitive to light for seed germination. They will germinate in light or in darkness. However, reclamationists are more concerned with grasses and many of these require light to germinate. For instance, Bermuda grass and Kentucky bluegrass seeds germinate more quickly in light. They should be in contact with a firm soil and mulched, but they should not be covered to the extent that all light is excluded.

After the seeds have germinated, the new seedlings must be able to break through any soil covering and into the open. The seeds may be prevented from breaking through if the soil has been compacted over the seeds, if a hard soil crust has formed over the seeds, if too much mulch has been applied, or if the seeds were planted too deeply. In most cases, larger seeds can be planted more deeply than smaller seeds and still obtain emergence (Appendix A).

There is not much danger of packing the soil too tightly over the seeds when they are broadcast and a drag of some type is used to cover them. However, there is danger of dragging too much soil over the seeds if the soil is loose and the soil surface is uneven. Of course, the use of a cultipacker after dragging may pack some soils too tightly, but this is not likely. Finally, there is always the danger of a soil forming a surface crust that is impenetrable to the new seedling. These crusts are more likely to form if the seedbed is extremely loose and porous. Therefore, the reclamationist should not overwork the seedbed and reduce soil aggregates to extremely small granules. Soils that have more than 70 percent sand and soils low in organic matter are more likely to form a crust. Fortunately the straw mulch used by many reclamationists helps to prevent rainfall from destroying the seedbed structure and forming these crusts.

It is known that the soil below the seeds should be packed with a pressure of approximately 5 to 10 $lb/in.^2$ to facilitate water movement through the soil and to the germinating seeds. However, this same pressure on the soil covering the seeds could prevent the seedling from emerging. Under these circumstances, an ideal planting procedure would be to pack the soil under the seeds, press the seeds into the compacted soil, and keep the soil loose directly above the seeds. It is also important to have a firm soil under the seeds in the event that the emerging seedling has to push through a compacted or crusted soil. If the seedling root has a firm soil to push against, many seedlings can break through a relatively strong crust. Equipment and practices approaching this ideal will better the chances for successful seed germination and seedling emergence.

BROADCAST SEEDERS

Many of the minesoils that require seeding are stony, have an uneven topography, or are steep. At any rate, they are unsuitable for sowing with a grain drill. In cases such as these a broadcast seeder of some type is required. These seeders throw the seeds onto the minesoil surface and a second operation may be required to cover the seeds. This second operation may be done with a harrow of some type, usually a type of tooth harrow, or a drag. If the seeds require light for germination, a straw mulch may be used to cover the seeds instead of soil. A better seedbed is obtained if a cultipacker is used before the seeds are broadcast. Compacting after sowing does not appear to be of significant benefit. Broadcast seeding is not as uniform or precise as drilling. The seeds are scattered in a random fashion, may be covered with too much soil, and fewer of the seeds are likely to produce a seedling.

The centrifugal broadcast seeder is probably the most popular of the broadcast seeders (Figure 34, and see Chapter 8, Figure 28). The seeder in Figure 34 can also be used to spread fertilizer and will hold from 100 to 2000 pounds of seed or fertilizer. Material is metered from the hopper

Figure 34. Centrifugal broadcast seeder mounted behind farm tractor.

to the distributing disc by means of a calibrated gate or a fluted wheel. The distributing disc rotates parallel to the soil surface and throws the seed from 20 to 40 feet depending on seed characteristics.

Another type of broadcast seeder called a field distributor resembles a fertilizer distributor and can be used to distribute fertilizer as well as broadcast seeds. This seeder does not have furrow openers or seed coverers. It consists of a long box hopper that is pulled with the long axis at right angles to the tractor path. The seed is metered from the storage hopper through an adjustable gate at the bottom of the hopper onto the soil surface.

AERIAL SEEDERS

Seeds, fertilizers, and pesticides can be quickly broadcast over large areas by fixed-wing aircraft or helicopters. This system has been used where the terrain is too rough for ground equipment or where the seeding has to be done within a short time. Also this system has been used frequently to sow forest seeds and to fertilize forests.

Fixed-wing aircraft distribute the seed by means of a centrifugal broadcast seeder or by means of a venturi-type spreader. Both of these systems cast the seeds from beneath the fuselage. The seeds are carried in a hopper inside the fuselage and move from the hopper to the spreader through a sliding gate controlled by the pilot. The centrifugal spreaders are powered hydraulically or electrically. The venturi spreaders use the propeller slipstream to move the seeds backward and to the side of the plane.

Both spreaders can be calibrated for any desired application rate. The seeds or fertilizer are spread in parallel flight paths. Markers on the ground are used to guide the pilot along the parallel paths. Several methods have been used to mark these paths. Two popular methods are long bamboo poles with colored ribbons at top or large colored balloons filled with hydrogen or helium and attached to long cords. These markers can be moved as the aircraft moves on to the next flight path. Pilots usually try to fly as low as possible to avoid seeds drifting and to obtain the best seed placement.

Broadcasting seeds from the air has all the disadvantages of any broadcasting system. The broadcasting is likely to be more successful if some method of covering the seeds with soil or mulch follows broadcasting. If the work is done on a windy day, the seeds may drift away from the desired site and may tend to concentrate rather than spread evenly over the soil surface.

Helicopters use centrifugal broadcast seeders attached to the sides of the helicopter or hanging beneath the helicopter (see Chapter 6, Figure 20). The rotating discs that broadcast the seeds can be powered hydraulically by the helicopter engine or they can be powered by a small gasoline engine mounted on the broadcaster. Seeding rate is determined by the size of the opening between the seed hopper and the rotating disc.

The costs of operating helicopters are greater than for fixed-wing aircraft, but the helicopter does more work in a shorter time because it does not require the longer travel time to an airstrip. The helicopter can be refueled or reloaded with seeds from a truck at the planting site. The helicopter also requires less turning and loading time and is more maneuverable at low speeds than a fixed-wing aircraft.

GRAIN DRILLS

Grain drills should be used when conditions will allow. These drills place the seeds at the proper depth, space them properly, cover the seeds and firm the soil around the seeds when necessary. This means that the seeds will have the opportunity to absorb needed moisture, but will not be planted too deeply to emerge from the soil surface. It also means that the seeds can be planted at the optimum spacing for best vegetative soil surface coverage with the least amount of seeds. Grain drills can also be fitted with fertilizer and grass seed hoppers. Therefore, grain seeds, grass seeds, and fertilizer can all be applied at the same time.

There are two general types of grain drills: end-wheel drills and press-wheel drills. The end-wheel drill has supporting wheels at the ends of the seed and fertilizer hopper to support the drill and drive a seed metering device (Figure 35). This drill opens a planting trench to the proper depth and deposits the seeds in the trench. The usual method of closing the trench is to allow loose soil to fall back into the trench under the influence of gravity or to have short sections of chain dragging behind each seed tube to push loose soil into the trench. If the soil needs firming, spring-loaded press-wheels can be attached to firm the soil covering each trench.

Figure 35. End-wheel seed drill with press wheels to firm soil over seeds.

Figure 36. Seed falls from metering device through the flexible tubing, between the discs, and into the furrow formed by the double discs.

Press-wheel drills have support wheels in front of the seed and fertilizer hopper to support the front of the drill, whereas the press-wheels support the rear of the drill, firm the soil around the seeds, and drive the seed metering device. There are several different types of press-wheels to be used for different soil conditions and row spacings.

The planting furrow for end-wheel or press-wheel drills may be opened by a small shovel plow or disc. The shovel opener is probably best for stony soils or soils that are crusty. These shovels can be spring-loaded to prevent damage when the shovel strikes a stone or immovable object. There are several types of shovels that can be used for different planting conditions.

The disc furrow-opener can be a single disc or a double disc. Seeds are deposited behind the single-disc opener and between the discs of the double-disc opener (Figure 36). Seeds can be planted to a precise depth with the double-disc opener when equipped with depth regulators and can be planted at high speeds. However, the seedbed should be well prepared, smooth, and nonstony for best performance.

GRASS SEEDERS

The grass seeder is a seed and fertilizer hopper mounted between two cultipackers. The first cultipacker smooths and firms the seedbed, creating small grooves in the soil. Seeds are broadcast from the seed hopper onto the prepared soil. The second cultipacker rolls over the seeds forming new grooves between the grooves made by the first cultipacker and thereby covering the seeds. Seeds are planted at the proper depth and in a firm seedbed. The rows of seeds are planted closely together and, under good weather and soil conditions, a dense stand of grass is produced. Grass seeders can be used to fertilize and sow seeds in one operation. These seeders do not operate well on rough seedbeds or stony areas. The second cultipacker could be removed if seeds requiring light are planted. A straw mulch can be applied after the planting operation to keep the soil surface moist for seed germination, hold the seeds in place, and reduce animal depredation.

RANGELAND DRILLS

The rangeland drill was developed to plant grain or grass seeds under poor seedbed conditions. Like other drills, it will apply fertilizer as well as seeds to the seedbed. It can be used to break crusted surface soils in the planting process. The rangeland drill is actually a heavy-duty, side-wheel grain and grass seed drill. It works well in rough and stony situations. The single-disc soil openers are independently suspended on trailing arms. If a disc strikes a stone or other obstacle, the disc rides up and over the obstacle while the other discs continue planting. Planting depth can be controlled by depth bands. Seed furrows are closed by trailing chains or pipe drags. The seed hoppers can be modified to handle a variety of seeds, including some tree seeds. Seeds are placed at a uniform depth and covered with soil. This drill should be operated on contour along slopes because it creates furrows that can become erosion gullies if the drill is operated up and down slopes.

HYDRAULIC SEEDERS

Hydraulic seeders, often called hydroseeders, can be used to apply seed, fertilizer, limestone, and mulch to a minesoil. These seeders consist of a tank that holds a slurry of water and the material to be applied to the soil (see Chapter 5, Figure 15). The tank has an agitator to keep the material suspended in the slurry. A pump pulls the water and suspended material from the tank and forces it out through a nozzle and onto the soil surface. Materials can be sprayed for distances up to 200 feet.

The discharge nozzle is mounted on top of the tank and will rotate 360° horizontally and 120° to 180° vertically. Material cannot be applied to an area within about 20 feet of the tank. Therefore the operator must fill in the missed areas as the hydroseeder moves forward. If necessary, a hose can be attached to the tank and the slurry can be applied through this hose to any area that cannot be satisfactorily reached by the nozzle mounted on top of the tank.

The slurry can hold 3 to 6 percent of solids by weight. Therefore, 0.25 to 0.50 pound of solids can be applied for every gallon of liquid. These solids can be seed, short fiber cellulose mulch, undissolved fertilizers, or limestone. Fertilizers that dissolve in the water would not be considered solids. Care should be taken in mixing materials in the slurry. Much of the nitrogen in fertilizers may be lost as ammonia gas if limestone or some other alkaline agent is placed in the tank at the same time.

Finally, the effectiveness of legume-inoculating bacteria will be drastically reduced if the reaction of the slurry is allowed to fall below a pH of 5.0.

These seeders can be used to seed steep slopes and other inaccessible areas. It is a fast and efficient method of applying seeds, soil amendments, and mulch in one operation. However, hydraulic seeders require large amounts of water, and extra amounts of seeds must be used when seeds and mulch are applied at the same time. These extra seeds are applied to offset the seeds lost due to their falling on top of mulch instead of the soil surface. Also, the agitators and pumps may damage many seeds. Recirculating the slurry through the centrifugal pump for as long as 30 minutes can reduce seed germination by a significant amount.

PLANTING TREE SEEDS

Practically all the tree seeds planted in mine reclamation have been pine and black locust seeds. Tree seeds have been planted by hand broadcasting, hand-operated broadcast seeders, powered broadcast seeders attached to tractors, and by broadcasting from aircraft. All of these methods have been used successfully with pine seeds in the Alabama coal fields. Broadcasting has been successful with black locust seeds in states farther north. Federal reclamation regulations will probably eliminate this method of reforestation. It is difficult to establish seedlings from broadcast tree seeds in a heavy cover of forage plants.

There are other methods of planting tree seeds such as the seed drill developed by the U.S. Forest Service (Figure 37). This implement can be used to remove competing vegetation, prepare a seedbed, and deposit the tree seeds on the soil surface. This implement could be used to plant tree seeds in a forage sod by attaching scalpers to remove competing plants from the immediate seeding area. All scalping should be done on contour to prevent excessive soil erosion.

Herbicides that kill grasses and legumes could be used instead of scalping to prevent excessive competition from forage plants. The herbicide could be applied before planting the tree seeds or at the same time, depending on the type of herbicide used. Forage plants would be killed in strips or spots and the tree seeds planted in these strips or spots. The dead forage would act as mulch to help prevent erosion, hide the seeds from predators, and keep the soil surface moist for seed germination.

Existing agricultural seed drills could be used or modified for planting many types of tree seeds along with forage plant seeds. Most of the planting tubes would be used for forage seeds, but one or two of the tubes could be used to plant rows of tree seeds at predetermined distances between the rows of forage seeds. Some tree species should be able to compete successfully with some forage plants under these circumstances.

Appendix B shows some of the information needed to plant many of the tree seeds that might be used to vegetate surface mines in the eastern U.S. More complete information on these tree seeds and many others can be found in the book, *Seeds of Woody Plants in the United States,*

Figure 37. U.S. Forest Service tree seed drill. The coulter is immediately behind the tractor, followed by the scalping plow, bed-forming discs, seed furrow opener with seed hopper and metering device above, and packing-wheel at the rear.

Agriculture Handbook No. 450. This book, published by the U.S. Forest Service of the U.S. Department of Agriculture, gives detailed information on the pregermination treatments required by most tree seeds to germinate in a reasonably short period. Appendix B gives a brief description of the pregermination treatments required.

SEEDLING TRANSPLANTING

Some crops such as tobacco, tomatoes, shrub seedlings, and tree seedlings are grown from seeds in nursery beds and then transplanted to the fields, yards, and forests where they will grow to maturity. In the past, field crops such as tobacco, tomato, and cabbage were transplanted to the field by hand, but today implements towed behind tractors open a furrow, place seedlings in the furrow, close the furrow, and pack the soil about the seedling roots.

Reclamationists are more familiar with the operation of transplanting tree seedlings on minesoils. These seedlings are planted either by hand or by a mechanical transplanter. There are differences in degree of difficulty when planting tree seedlings on reclaimed surface mines. If the regraded overburden is not topsoiled it will usually be stony and difficult to hand plant or machine plant. This is not true of all overburden material, but it is true of most overburden. If the overburden has been topsoiled to a depth of 8 inches or more, there should be little hand or mechanical planting difficulty. An exception occurs when the topsoil has been excessively compacted. In this case the topsoil may have to be treated as overburden.

Stony or compacted soils are easier to plant if a subsoiler or ripper is pulled through the soil ahead of the hand or mechanical planter. Soil and other material fall back into the groove cut by the ripper, forming a stone-free vertical cut through the soil that can be easily planted with a dibble or mattock. If the soil within the range of the ripper contains large stones, some of these stones will be pulled to the surface and may interfere with the operation of a mechanical tree planter.

Seedling transplanters should not be used where slopes are greater than 20 percent or where large rocks are present. Most transplanters create a continuous furrow and care must be taken to prevent the furrows from becoming erosion channels. These and other precautions must be taken for any tree-planting operation. It is usually best to have a trained supervisor for every two machines. This supervisor can ensure that the machines are well supplied with seedlings and properly maintained. The supervisor can also make constant inspections for planting quality by removing the soil around the seedling roots to determine depth of planting and placement of roots. Spacing of the seedlings must be checked constantly. There is a tendency for personnel to change spacings as planting conditions or tractor speeds change. Spacings that are too close can be as bad as spacings too distant. Finally, the supervisor can assure that the seedlings are stored under the best available conditions from the time they leave the nursery until they are planted.

FORAGE SEED MIXTURES

There is usually at least a moderate amount of variation in soil characteristics within any reclaimed surface mine. Because of this it is almost always desirable to plant a mixture of forage plant species rather than a single species. When a mixture of species is planted, it is much more likely that the diverse soil and weather conditions will be satisfied and a complete stand of vegetation obtained.

At least one legume should be included in the seed mixture to help maintain the nitrogen content of the soil. Mixtures of grass and legume plants will tend to keep the surface of the soil covered with living vegetation during a greater part of the year than would a single species. A quickly-establishing annual may be desirable in the mixture in order to have the soil surface protected from erosion as soon as possible. These mixtures of plants will also establish a soil cover over a greater range of weather conditions than any single plant. It is wise to include at least one plant that can withstand minor droughts for those times when unexpected droughts occur during the period of establishment. It is probably best to plant only half or less of the recommended amount of seed for the quickly establishing annual. This is done to prevent excessive competition for the slower establishing perennials.

Appendix A is a listing of plants that can be used for reclamation in the eastern part of the U.S. Appendix A also indicates the regions in which these plants can be used and the time of the year when the seeds should be planted. The time of year for planting as shown in Appendix A is the one determined by agriculturists to be best to produce a mature crop. Reclamationists often do not have to produce a mature crop within a specified time. They may need only to produce a quick ground cover. Under these circumstances, planting can be done at any time that soil moisture and temperature will provide seed germination and plant establishment. The map shown in Appendix A identifies the planting regions. A mixture of seeds can be obtained for a particular mine by selecting one or more perennial grasses, one or more legumes, and one annual from Appendix A. The seeds selected should be ones that are adapted to the region in which the mine is located and the existing soil conditions.

CHAPTER 10

MULCHING

The most important time to have sufficient soil moisture and the correct temperature range at the soil surface is while seeds are germinating and the young seedlings are becoming established. Most grass seeds and other small seeds are planted at or close to the soil surface, and the soil surface is the soil layer most likely to dry or have wildly fluctuating temperatures. The cultural practice that can most likely maintain soil moisture and modify temperature extremes at the soil surface is mulching.

Mulches applied to newly seeded minesoils will help to control soil erosion, reduce the loss of moisture from the soil surface, modify soil temperatures, prevent soil crusting, increase the amount of water stored in the soil for plant use, reduce animal predation, and hold the newly planted seeds in place. Under ideal soil and weather conditions, mulching is not needed. Under extremely poor weather and soil conditions, mulching will not help to establish a stand. However, the vast majority of weather and soil conditions fall between these two extremes and mulching is invaluable under those conditions.

The advantages of mulching for young plants as well as for seeds should be kept in mind. The soil and climate modifications of mulching can be beneficial for transplants as well as for seeds and seedlings. Mulching shrub or tree seedlings may be worthwhile under adverse conditions to keep the seedlings alive until they become established.

There are many materials that can be used for mulches. However, the three most popular materials in reclamation have been straw, hay, and wood cellulose fiber. Practically all research work has shown that straw or hay is equal to or better than other materials for mulching. Other materials such as shredded bark can be used successfully when available.

There is good agreement among reclamationists and other workers that some rate between 1 and 2 ton/A straw or hay is best for seed germination and seedling establishment. Table 15 shows the effects of different straw mulch rates, slopes, and soil textures on soil erosion

Table 15. Effect of Straw Mulch Rate, Slope, and Soil Texture on Soil Erosion Loss After 2 Hours of Simulated Rainfall

Straw mulch rate (ton/A)	Percent soil surface covered	Loam (15)[1] (ton/A)	Silt loam A (5) (ton/A)	Silt loam B (3) (ton/A)	Silt A[2] (10) (ton/A)	Silt B[2] (8) (ton/A)
0.0	0	27.8	12.4	14.5	19.3	5.9
0.25	33	9.0	3.2	6.0	7.5	3.1
0.50	50	8.7	1.4	3.5	4.4	2.0
1.0	75	5.1	0.3	1.7	4.2	1.7
2.0	90	1.1	0.0	—	—	—
4.0	95	0.7	0.0	—	—	—

[1]Numbers in parentheses indicate percent of slope.
[2]Silts A and B were artificial soils of silt-sized particles.
Source: Meyer, L.D. 1970. "Mulch Rates Required for Erosion Control on Steep Slopes." Soil Science Society of America Proceedings 34:930. Used by permission.

losses. The most practical rate seems to be between 1 and 2 ton/A. However, there is an overall lessening of erosion as the slope steepness decreases. If a reclamationist wished to decrease the amount of mulch used, the amounts used on level areas and gentle slopes could be drastically reduced (0.25 to 0.50 ton/A). It is probably best to apply 1 ton/A or more on slopes of over 10 percent. The 1 ton/A would leave about 25 percent of the soil surface exposed. A reclamationist could probably make a valid ocular estimate of the amounts of straw being applied by noting amounts of soil surface covered or exposed, as noted in Table 15.

Mechanical mulch blowers are used to apply square bales of hay (see Chapter 5, Figure 14). These implements separate the stems, chop them into shorter lengths, and blow them through a nozzle that can be moved vertically and horizontally. Under good conditions the mulch may be blown as far as 70 feet. Every effort should be made to apply the mulch as uniformly as possible. If the mulch does not cover the soil surface adequately, seed in these thinly covered areas may not germinate and these areas may have inadequate or no vegetation. If the mulch is too thick, the seedlings may not break through into the light and, again, these areas may have inadequate or no vegetation. Overly thick mulches may also create conditions near the soil surface that interfere with seed germination and seedling establishment. Reclamationists in the more northern states should keep in mind that thick mulches slow the warming of soils in the spring. This could slow the germination and early growth of many grasses and legumes.

Straw or hay mulch can be held in place by several methods but the two most practical methods for large areas are spraying the applied mulch with emulsified asphalt or pushing parts of the mulch into the soil with discs or crimpers. The asphalt can be applied with special spraying equipment after the mulch has been applied, or some blowers have attachments that allow the asphalt to be blown out with the mulch in one operation. The recommended amounts of asphalt are between 200 and 800 gal/A, depending on the susceptibility of the mulch to movement by wind or water and gravity.

Any method of cutting or pressing the mulch into the soil will help to stabilize the mulch. The implements that have been used for this job are cutaway disc harrows, sheep's-foot rollers, brush harrows, and mulch packers. However, the best implement is probably a crimper. This implement presses part of the straw into the soil and leaves part protruding above the soil surface. The resulting situation is much like the stubble left on the soil after cutting hay.

The crimper consists of a series of 20-inch diameter flat discs spaced approximately 8 inches apart on a horizontal axle. Two steel boxes are attached to the frame holding the axle. These boxes can be filled with weights when needed. The crimper has rubber-tired wheels that can be hydraulically lowered for transportation. This implement should be operated on the contour to prevent the formation of grooves running up and down slopes that might encourage soil erosion. Caution should be exercised when operating in stony conditions, as with any disc implement.

Hay and straw are often baled in large round bales. A machine has been developed that picks up these large bales (1500 pounds), carries them to a feeding area, shreds the hay, and places it in windrows for cattle feeding. For mulching, a powered flail is added to the discharge nozzle. This flail strikes the straw that would ordinarily be placed in a windrow and distributes it over a much wider area. A uniform distribution of hay and straw is accomplished with this system, but it is not suited to rough or steep areas. After distribution, the mulch should be anchored by asphalt or crimping.

There is also a mulch spreader that will spread many types of material such as square bales, round bales, bark, and wood chips. This machine is a modified tub hay-grinder and must be loaded by a separate machine such as a front-end loader. However, it will carry and spread two large round bales in one load. Large round bales are spread at a rate of 1000 to 1500 pounds per 3 or 4 minutes. Another one of the newer mulch spreaders will load itself with round bales of hay or straw but does not function well with other bales or mulches. This machine will only carry one large round bale, but it also spreads these bales at the rate of 1000 to 1500 pounds per 3 or 4 minutes.

Manure spreaders have also been modified to spread straw or hay. These are wagon-type implements and probably should be used on level or gently sloping lands only.

The application of mulch is time-consuming and costly. However, it is an essential operation when soil moisture is not sufficient to keep the bare soil surface moist for seed germination and seedling establishment. Mulching may be thought of as insurance. There are times when it is not needed, but it is needed often enough to make its use economically reasonable.

CHAPTER 11

FORAGE STAND MAINTENANCE

The following discussion of fertilization is based on forage production for grazing or hay on unmined lands. It does not take into account any government regulations that may interfere with a fertilization program. If regulations prevent fertilization after initial plant establishment, steps can be taken to abide by these regulations. For instance, many soils and overburdens have a good supply of potassium. This means that nitrogen and phosphorous are probably the only nutrients that need be added, and if the forage stand contains at least 30 percent legumes there is usually no need to add nitrogen. Phosphorous tends to remain in the soil for relatively long periods of time. Therefore, it can often be added in sufficient quantities during initial fertilization to supply the plants for a long period of time. If regulations do prevent subsequent fertilization, it seems reasonable to apply extra amounts of phosphorous and potassium and to be sure that the forage stand contains at least 30 percent legumes. Stands of *Sericea lespedeza* in Alabama have been known to provide 100 percent soil cover for 10 years or longer when adequate amounts of phosphorous and potassium were present in the vegetated overburden.

One of the principal objectives of every reclamationist is the establishment of a forage stand that protects the minesoil and produces a crop for harvest. Most reclamationists do well insofar as establishment is concerned, but they tend to neglect the maintenance of the forage stand after establishment. If the stand is not maintained it may deteriorate rapidly and not give adequate soil protection. Also, if the initial stand does not meet soil cover standards, there is a good possibility that an adequate maintenance program can increase the vegetative soil cover without resorting to replanting.

The maintenance practice most often needed is fertilization, and the nutrient most often needed is nitrogen. The forage stand should be inspected periodically during the growing season for growth, disease, insect, nutrient deficiency, and mineral toxicity problems. Disease symptoms may appear in the roots, stems, or leaves. A plant that is dying with no

Table 16. Possible Symptoms Appearing as Grasses Become Increasingly Deficient in a Specific Nutrient

Nutrient	Deficiency symptoms	First appearance
Nitrogen (N)[1]	Overall light green or yellow-green color. Leaf begins to die at tip.	Older leaves
Phosphorous (P)	Blue-green color becoming purple as the deficiency increases. Most often seen on young seedlings.	Older leaves
Potassium (K)	Leaves develop yellow streaks and tips and margins become brown as deficiency increases.	Older leaves
Calcium (Ca)	Reddish-brown color develops between leaf veins. Growing tips die and curl as deficiency increases.	Newer leaves
Magnesium (Mg)	Leaves develop unusual green or yellow-green stripes that change to a bright red.	Older leaves
Sulfur (S)	Leaves become yellow and leaf tips die as deficiency increases.	Newer leaves
Copper (Cu)	Plant stunted, leaves yellow, leaf tips turn brown, stem tips die.	Newer leaves and stem tips
Molybdenum (Mo)	Leaves pale yellow, leaves wither.	Unknown
Iron (Fe)	Rapidly increasing chlorosis between leaf veins, leaves become white in severe cases.	Newer leaves
Zinc (Zn)	Leaves become small, crowd together and may be yellow or have white stripes.	Newer leaves
Manganese (Mn)	Younger leaves become yellow and yellow-green spots on older leaves, leaf tips wither.	Newer leaves
Boron (B)	Leaf tips pale green and stem tips die.	Newer leaves and stem tips

[1]Chemical symbol given in parentheses.

unusual leaf discoloration probably has a root disease or is being attacked by a root-chewing insect. Other disease and insect problems will usually be obvious, and an expert in these fields should be called in for consultation. In most cases growth and nutrient deficiency problems can be handled by the reclamationist.

If a nutrient is only slightly deficient or even moderately deficient, the only symptom is likely to be a decrease in plant growth. If the forage does not appear to be growing as fast as it should, the cause is likely to be nitrogen deficiency. However, this lack of growth can be due to the deficiency of some other nutrient, abnormal weather, or poor soil condition.

Soil analysis will probably reveal any nutrient deficiency, and the reclamationist will usually be aware of any abnormal weather condition such as drought or unusually cold weather. Also, there are characteristic symptoms for each nutrient that can be noted when a nutrient becomes increasingly deficient. Table 16 shows some characteristic symptoms for each of the nutrient deficiencies of grasses. These symptoms may sometimes confuse the observer, especially when more than one nutrient is deficient. Also remember that the first deficiency symptom is usually a lack of growth, and this is not as obvious as a change in leaf color.

Mineral toxicities also cause changes in the appearance of a plant. For instance, excessive amounts of copper, manganese, zinc, chromium, cadmium, cobalt, or nickel in the soil may cause iron deficiency symptoms in the plant. This could obviously lead to confusion. Because of this it is usually worthwhile to obtain the advice of an agronomist or soil scientist when unusual plant symptoms appear. These scientists will use some combination of visual symptoms, plant tissue analysis, soil analysis, and past experience to solve the problem.

NITROGEN FERTILIZATION

A nitrogen deficiency may occur at any time in the life of a plant and usually can be satisfactorily corrected when it occurs. The use of nitrogen fertilizers in a forage stand depends on the plant composition of the stand. The stand may be composed of grasses alone, grasses and legumes, or legumes alone. If grasses are grown without legumes, nitrogen will almost certainly be needed at the time of sowing, and a second application during the first growing season will be beneficial. Grasses will also benefit from nitrogen applied at least once during each succeeding growing season. Grasses will respond favorably to nitrogen applied as often as every 30 days during the growing season if there is adequate soil moisture. However, this frequent fertilization is not necessary unless maximum forage production is desired for grazing or hay. The amount of nitrogen applied after establishment should be at least half of the amount recommended by the soil testing laboratory at the time of sowing.

A forage stand composed of grasses and legumes has some advantages over one composed of grasses alone. The foremost advantage in mine reclamation is the decreased cost of nitrogen fertilization. After a legume becomes established and is functioning properly, it will produce enough nitrogen for its own growth plus a surplus that can be used by the grasses. Also, the legume may grow well on parts of the minesoil where grasses would not grow as well. Finally, the legumes make the forage more palatable and nutritious for grazing animals or hay.

Nitrogen fertilizer is usually applied to a soil planted with grass and legume seeds to carry the seedlings through the establishment period. As a rule of thumb you do not need to add nitrogen to an established forage stand if it contains more than 30 percent legumes. If the legume percent is 30 to 20, apply 30 pounds of nitrogen per acre in the spring for summer grasses and in the fall for winter grasses. If the legume component is less than 20 percent, apply 30 pounds of nitrogen per acre in the spring and in the fall. It is believed that the addition of relatively large amounts of nitrogen fertilizer to legumes causes the nitrogen fixing bacteria in the legumes root to cease nitrogen fixation. Therefore, such fertilization is considered wasteful. Although legumes do not require nitrogen fertilization, they usually require more phosphorous, potassium, and lime than the grasses. Legumes do not seem to extract nutrients from the soil as easily as grasses, and most legumes require a neutral or only slightly acid soil for best growth.

PHOSPHOROUS FERTILIZATION

Phosphorous deficiency is more likely to appear in young forage stands immediately after establishment than in older stands. Soil testing and fertilizer recommendations will prevent such a problem. To produce an adequate stand a phosphorous fertilizer should be added immediately if the need is suspected. If the phosphorous application is delayed the stand may never develop adequately.

Phosphorous will accumulate in soil to a greater extent than nitrogen or potassium. Therefore, the need for phosphorous in a minesoil may be great in the first years of vegetation but is likely to decrease with time. The need for phosphorous in the first years of establishment is likely to be higher in alkaline or very acid soils. Both of these soils will convert much of the added phosphorous to an unavailable form. Phosphorous fertilizers are usually applied to forage crops in the spring or fall. The amount applied should be determined by a soil testing laboratory in order to apply adequate amounts but avoid detrimental accumulations in the minesoil.

POTASSIUM FERTILIZATION

Potassium should be applied to the soil before the forage plants begin growth in the spring. It has been shown that adequate soil potassium will cause grasses and legumes to produce earlier spring growth. Grasses tend to take up potassium more efficiently than legumes and will utilize potassium at the expense of the legumes. Potassium appears to be particularly important in obtaining regrowth after a forage stand has been harvested or heavily grazed.

When grasses and legumes are grown together, the amounts of needed potassium should be determined by a soil testing laboratory. If the laboratory is not told that grasses and legumes will be grown together, the potassium recommended by the laboratory should be increased by approximately 10 percent in order to satisfy the legume needs.

CALCIUM AND MAGNESIUM AMENDMENTS

Calcium can be supplied to a minesoil as calcitic limestone. Both calcium and magnesium can be supplied to a minesoil as dolomitic limestone. These limestones should be applied and worked into the minesoil before sowing. Legumes will usually make their best growth at a pH of about 6.5, if other factors are not limiting. Also, legumes are more sensitive than grasses to a lack of soil calcium. Many minesoils, especially an acid overburden, will inactivate limestone within a short time. Most legumes need this lime and will exist only for a short time in a forage stand if the minesoil is not properly limed. A need for limestone can be detected by periodic soil tests, and more limestone can be worked into the soil profile to restore the pH to the desired level.

OTHER NUTRIENTS

Plants use only small amounts of the other nutrients and most minesoils will contain adequate amounts. Micronutrients should not be applied unless the need has been determined by soil analysis. An excess of many of these nutrients can be as bad or worse than a deficiency. Certain soils and conditions are more likely than others to produce micronutrient deficiencies. Table 17 describes some of these soils and conditions, and the type of plant most likely to be affected.

FORAGE PRODUCTION FOR GRAZING

The previous discussion of forage fertilization was concerned with producing an adequate soil cover as quickly as possible. There was no concern for the production of maximum amounts of forage for grazing or hay with the greatest economy. However, the only additional need is to increase the amounts of fertilizer to produce greater amounts of forage. Of course, fertilizer will not be effective if some other factor such as water is limiting.

If the forage is to be used for grazing in the winter, it should be fertilized as needed with phosphorous and potassium in the late summer or early fall. If the forage is to be grazed or cut for hay in the summer, it should be fertilized in the early spring. Grasses should receive about 50 pounds of nitrogen per acre plus any needed phosphorous and potassium as determined by soil testing. If legumes are growing with the grass, the forage probably will not need nitrogen during the period when the legumes are active. Nitrogen should be added to the forage when the legumes become less active. This usually means adding nitrogen in middle or late summer.

Table 17. Soil Conditions Most Likely to Produce Micronutrient Deficiencies and Plants Most Likely to Show Deficiencies

Micronutrient	Soil condition	Plant
Boron (B)[1]	Dry, sandy soil, very acid soil, very low pH, overlimed soil, high pH	Legumes
Copper (Cu)	Sandy soil, overlimed soil, high pH	Grasses
Iron (Fe)	Alkaline soils, excess phosphorous, high pH	Grasses
Manganese (Mn)	Sandy soils, alkaline and calcareous soils, high pH	Legumes
Molybdenum (Mo)	Acid soils, leached soils, low pH	Legumes
Zinc (Zn)	Sandy soils, calcareous soils, leached soils, excess phosphorous, high pH	Grasses

[1]Chemical symbol given in parentheses.

FORAGE CULTURE SUMMARY

There are many factors that must be taken into account in the establishment of a forage stand. If one or more of these factors are neglected by the reclamationist, the chances for success are greatly reduced. The factors can be grouped under three general headings for the purpose of a quick consideration of them all. Each of the factors has been examined in more detail in other portions of this book. The three general headings are germination, establishment, and growth.

I. Germination
   A. Seed quality—The seeds must be alive when they are planted. This means that live seeds must be acquired and be properly stored until planted. Also, the seeds must not be in a dormant or hard-to-germinate stage if immediate germination is desired.
   B. Moisture—There must be enough soil moisture for seed germination. Alternating moist and very dry conditions can destroy seed viability.
   C. Temperature—The reclamationist usually controls this factor by planting during the proper season. However, unseasonable temperatures may still prevent germination.
   D. Soil air—Wet soil may not allow enough air to reach seed for germination, but this does not often happen.

E. Mulch—Mulch that is too thin may not keep the seed moist enough for germination. Mulch that is too thick may prevent the seed from warming enough to germinate.

II. Establishment

A. Depth of planting—Seeds that are planted too deeply cannot emerge from the soil surface. This is probably the factor that causes more forage plant loss than any other factor when soil moisture is adequate.

B. Drought—A light rain may germinate seeds but not keep the soil moist long enough for seedling establishment. Mulches help prevent this situation.

C. Freezing—Newly germinated seedlings are sensitive to freezing. A temperature below 26°F will probably kill seedlings. Therefore, any planting that is likely to subject seedlings to these low temperatures is to be avoided.

D. Soil crusting—Some soil textures are likely to crust if they are overtilled or if organic matter is low. These soil crusts prevent the seeds from pushing through to the soil surface after germination.

E. Mulch—Mulch of the proper thickness will retain moisture at the soil surface for seedling establishment. Thick mulch may prevent the emergence of seedlings. Mulches help to prevent soil crusting.

III. Growth

A. Drought—This is probably the most common reason for lack of forage growth.

B. Poor soil drainage—Excessive water in the soil will exclude soil air and forage plants will lose growth or even die.

C. Soil nutrients—In grasses, nitrogen is the nutrient most likely to be lacking, as is phosphorous in legumes.

D. Soil pH—The soil must be within the optimum pH range for best growth of the plant or group of plants being grown on the minesoil.

E. Insects and diseases—The forage crop must be examined frequently to detect these problems for timely control.

F. Cold weather losses—Species must be planted that are adapted to the prevailing weather. New crops must be allowed to establish completely before cutting or grazing, and they must be allowed to establish well enough to withstand frost heaving.

G. Legume inoculation—Legumes may die after germination if there is a lack of soil nitrogen and the proper bacteria are not present for root inoculation. The reclamationist should make an effort to ensure that all legume seeds are inoculated with the proper live bacteria and that the bacteria are placed under conditions that will keep them alive until they can penetrate the legume root.

CHAPTER 12

PRIME FARMLAND

The principles and regulations that apply to reclamation in the rest of this manual also apply to prime farmland reclamation. However, instead of placing the A horizon or a mixture of A and other horizons over the overburden, the A and B soil horizons must be placed in sequence over the overburden, and the reclaimed soil must be as productive as the original soil. The A horizon is usually defined as the mineral horizon at or near the surface in which an accumulation of decayed organic matter is mixed with the mineral material. The A horizon can also be the plowed surface soil which might originally have been part of the B horizon. The B horizon is the mineral horizon below the A horizon. The purpose of this type of reclamation is to preserve land that may be vital for future crop production. This purpose is a worthy one, but the procedure prescribed by regulations is probably not best in all cases and possibly not best in most cases. Many soils have horizons that are not good for plant growth and may even be toxic. It seems reasonable to identify these horizons and mix them with some better horizon to lessen their effect or not use them in the reconstructed soil. It is also likely that horizons of varying physical characteristics and varying chemical characteristics could be mixed to produce a horizon superior to the original horizons. Present regulations state that exceptions to the regulations will be granted if research data indicate that such exceptions will result in equal or better crop production. Under these circumstances it is probably worthwhile for mining companies to employ the services of a soil scientist to identify such reclamation options. The scientist can examine the soil profile before mining begins and provide the mining company with information regarding possible beneficial alternatives to complete and separate horizon replacement.

The first information needed for prime farmland reclamation is the location of these prime farmland areas within the mining permit. This can be done by U.S. Soil Conservation Service personnel, and is usually done in the process of obtaining a mining permit. If the prime farmland occurs in small, separated areas within the permitted area, it would probably be best to consolidate the small areas into one larger area at the time of reclamation. This would significantly increase the value of the land for agriculture purposes. However, such action would probably have to be approved by the regulatory authority.

SOIL MOVING

The soil horizons or horizon mixtures may be stockpiled in separate areas for future replacement or they may be removed from their original position and carried directly to their final position on the reclaimed area. In either case the soil will tend to be compacted in the processes of removal, stockpiling, and replacement. This is unfortunate but difficult to avoid when wheeled vehicles, such as pan scrapers, are used and they pass repeatedly over the reclaimed area. Compacted soils have smaller amounts of available water and air, and roots do not penetrate the soil as well as in less compacted soils. All of this means that the roots' ability to obtain nutrients and water is restricted, and this reduces the plants' ability to grow.

Pan scrapers are probably the equipment most extensively used for prime farmland soil hauling. They are versatile, flexible, maneuverable, and readily available from equipment distributors. They are capable of removing the soil from its original position, carrying the soil to its final position or a stockpile, and spreading the soil over the area to be reclaimed. The only assistance needed in these operations is pushing or pulling the pans by another vehicle when removing some soils.

There are alternatives to pan scrapers such as dump trucks. However, dump trucks cannot operate under the diverse conditions under which pan scrapers can operate, and dump trucks require additional equipment to pick up the soil material and load it onto the trucks. The advantage of the dump trucks is that they do not compact the soil as much as pan scrapers. Low ground-pressure crawlers are used to spread and level the soil after the dump trucks have delivered the soil.

Conveyor belt systems are another possibility for moving soil from the point of removal to the point of deposition. Such a system requires careful engineering, a mine with a long life expectancy, and a reasonably level topography. The high initial cost and lack of flexibility also make conveyor belt systems unattractive for most mines.

A piece of equipment with the maneuverability of the pan scraper and the rear-end dumping capability of the dump truck would be desirable. This equipment would have to be self-loading, articulated, all-wheel drive, and have a low ground-pressure. It is doubtful that such a machine will be developed for the single purpose of reclamation, but it is possible that equipment more desirable than the pan scraper or dump truck will be developed.

SOIL COMPACTION DETERMINATION

The soil can be examined for compaction after replacement and final grading. A fast method of determining the degree of soil compaction is to make bulk density determinations as described in Chapter 2. This can be done immediately after the soil has been graded. Root growth usually becomes restricted when the bulk density reaches about 1.4 g/cm^3 in fine-textured soils and about 1.6 g/cm^3 in coarse-textured soils. However, the only sure way to determine the effect of compaction is to grow plants on the soil and observe the root penetration that occurs in the soil. Roots will stop moving downward or the number of roots will be drastically reduced when they reach a compacted soil layer. The same thing will happen if a layer of soil is too wet and not aerated or if the layer of soil is toxic. Do not confuse these problems with compaction.

Another instrument used to determine a root's ability to penetrate a soil is the cone penetrometer. Using this instrument is the fastest method of locating compacted soil layers, but it requires experience and there are several sources of error. Reclamationists can learn to use this instrument successfully in their own areas with time and practice.

REDUCING SOIL COMPACTION

The method of correcting soil compaction requires much consideration on the part of the reclamationist. The purpose of correcting a compaction problem is to increase the volume of soil that can be utilized by roots and increase the amount of water and nutrients available to the roots.

If the compaction is occurring in a thin layer where horizons meet, it is possible that it could be corrected by plowing, discing, or harrowing before the next horizon is applied. Of course, this compacted layer may also be broken by using a subsoiler or ripper after all the horizons have been replaced. Such decisions must be made on the basis of cost, equipment availability, and effectiveness of treatment.

Compacted soil layers near the soil surface (6 to 12 inches) can be loosened with plows, discs, or harrows. Deeper layers (2 to 4 feet) can be loosened with subsoilers or rippers. It should be kept in mind that all soil manipulation should be done near the optimum moisture content. Subsoiling or plowing soils that are too wet is ineffective and often detrimental. Some dry soils can be subsoiled successfully. New equipment for soil tillage is being constantly developed. A German company has developed a machine that is designed to loosen compacted soil layers to depths of 4 feet without as much soil disruption as that of a conventional ripper. Soil tillage methods that mix soil horizons after replacement must be considered carefully. Such mixing may be unacceptable to the regulatory agency and should be cleared with the agency before any such field operations. Each piece of equipment must be evaluated by the reclamationist in view of the particular set of circumstances.

The depth to which a soil is compacted is important. If the soil is compacted below an approximate depth of 6 inches, some mechanical method of loosening the soil will probably be beneficial. However, many compacted soils are loosened in the surface region by freezing and thawing, wetting and drying, and root growth. These actions can greatly reduce surface soil compaction within 1 year.

REVEGETATING PRIME FARMLAND

All of the principles previously discussed regarding seedbed preparation also apply to prime farmland. These soils should be sampled and analyzed for limestone and fertilizer recommendations. This should be done as quickly as possible on these erodible soils. After the soil amendments have been mixed with the surface 6 to 12 inches of soil, the necessary type of seedbed should be prepared for the crop to be grown.

It is beyond the scope of this manual to discuss all possible crops that might be grown on prime farmland. Information regarding any desired crop can be obtained from county agriculture agents, agronomic consultants, and state agriculture schools. These agencies are invaluable when deciding which crops are adapted to a soil with specific characteristics and in a specific climate. Some crops produce root systems that penetrate compacted soils better than other crops. It is probably worthwhile to use these crops in the first rotation to improve soil structure for future vegetation production.

SUGGESTED READING: PART II

Finck, A. 1982. *Fertilizers and Fertilization.* Deerfield Beach, FL: Verlag Chemie.

Kepner, R.A., Roy Bainer, and E.L. Barger. 1978. *Principles of Farm Machinery,* 3rd ed. Westport, CT: Avi Publishing.

Martin, J.H., W.H. Leonard, and D.L. Stamp. 1976. *Principles of Field Crop Production.* New York: Macmillan.

U.S. Department of Agriculture, Forest Service. 1974. *Seeds of Woody Plants in the United States.* Agriculture Handbook No. 450. Washington, D.C.: U.S. Government Printing Office.

Vogel, W.G. 1981. *A Guide for Revegetating Coal Minesoils in the Eastern United States.* General Technical Report NE-68, U.S. Department of Agriculture, Northeastern Forest Experiment Station, Broomall, PA.

PART III

SOIL EROSION AND SEDIMENTATION

INTRODUCTION

Soil erosion cannot be totally eliminated. Indeed, some types of slow erosion help to build soils by reducing stone to soil-sized particles over long periods. This soil building is natural and probably in balance with natural soil destruction forces. Humans often interfere with this rate of change through activities such as surface mining. This greatly speeded erosion caused by humans must be controlled in such a way that the soil losses are minimized and the harm to surrounding land and water areas eliminated. Fortunately there are methods of water-handling that can give adequate erosion control in all but the most violent storms. It should be noted that unmined areas are also eroded by these violent storms.

It is unfortunate that complete control cannot be attained, but some price must be paid for unusual activities by humans or by nature that disturb a natural balance. It has been estimated that 4 billion tons of soil are lost in the U.S. each year due to water erosion. About 3 billion tons are lost through farming and forestry activities, and the other 1 billion tons are lost through all other activities. Much of this loss from human activities could be avoided, but not all.

There are two general long-term methods of preventing or controlling soil erosion from rainfall. These two methods are the building of mechanical structures and covering the soil with live vegetation. Covering the soil with live vegetation controls erosion through the following actions:

1. Vegetation prevents raindrops from striking the soil where they would move soil particles and begin the erosion process.
2. Vegetation near the soil surface prevents or slows water running over the soil surface. This reduction in velocity of water movement reduces the amount of soil that can be moved by the flowing water.
3. Vegetation removes water from the soil, thereby making room for the absorption of more rainfall by the soil. Water absorbed by the soil is not available for downhill flow.
4. Vegetation roots form channels in the soil for water infiltration into the soil. Dead vegetation in the soil provides food for earthworms, insects, and other animals that also form soil channels and improve soil permeability to water and air.
5. The residue from dead vegetation promotes good soil structure that allows water to move into the soil and retain the water.
6. Vegetation roots bind soil particles together to make the soil more resistant to erosion and promote soil structure.

Mechanical structures control erosion by the following actions:

1. Runoff water can be diverted from a slope into a channel that carries the water over a lesser slope to a safe discharge point.
2. Water passing over a slope can be intercepted at intervals by a level area to decrease the water velocity and thereby remove some sediment from the water.
3. Impervious soils can be loosened by plowing to allow more rainfall to infiltrate the soil rather than running over the soil surface.
4. Excess water that would move downhill quickly can be held in terraces with closed ends or ponds until the water infiltrates the soil or is passed downhill over a long period of time and in small quantities.
5. Erodible materials can be covered with a nonerodible material such as stone.

This section of the manual is concerned principally with these mechanical methods of erosion and sediment control.

CHAPTER 13

SOIL EROSION FROM RAINFALL

A single raindrop does not have a great effect on soil, but when all the raindrops in an inch of rain are combined, they weigh 226,500 pounds per acre. These raindrops fall at a velocity of about 25 feet per second. This mass and velocity result in a tremendous amount of energy applied to each acre during a rainstorm and can cause a great amount of soil damage. Falling raindrops have several detrimental effects on soils. They tend to break soil aggregates into smaller aggregates and even into individual soil particles. These smaller aggregates and particles are easier for water to carry downhill. Raindrops also tend to compact soils and wash fine soil materials into the pores of the upper few inches of soil. These two factors seriously reduce the number and size of soil pores, which results in less water infiltrating the soil. Any water that does not infiltrate the soil will run over the soil surface and is likely to carry soil particles downhill. A lack of infiltration also means that less water will be stored by the soil to be used for plant growth.

The above paragraph reveals that raindrops alone have much to do with soil erosion. However, what about the combined raindrops after they reach the ground and move over the soil surface as a sheet or stream of water?

SHEET EROSION

Sheet erosion is the uniform removal of soil particles from large areas of the soil surface by flowing water (Figure 38). This takes place after the soil particles have been dislodged from the soil surface by raindrops. This uniform erosion of soil is not as noticeable as rill or gully erosion. Therefore, it is more dangerous than other forms of soil erosion, as large amounts of soil may be removed before the erosion is detected. Sheet erosion is more detrimental on natural soils than on minesoils. On natural soil the organic matter and nutrients are likely to be concentrated in the soil surface, where they are the first to be lost in sheet erosion. On mine-soil the surface soil is likely to be well mixed and any surface soil lost

Figure 38. Sheet erosion has taken place on this soil. Note the pebbles left on pedestals after raindrops have dislodged adjoining soil particles and washed them downhill.

to erosion will be no better than the remaining soil. Also, the lost soil can probably be replaced from topsoil storage.

RILL EROSION

The soil surface is seldom uniform in topography and water running downhill tends to concentrate in small channels. These channels become deeper and noticeable within a short time after erosion begins. When these channels are small and can be eliminated by normal tillage operations, they are called rills (Figure 39). When the channels can no longer be eliminated by normal tillage, they are classified as gullies (Figure 39).

GULLY EROSION

Gullies form through more than one process. The most common process is when a rill gradually deepens and widens due to the increasing amount of water flowing down the rill during each succeeding rainstorm. A second process is the fall of water over a ridge. In this process the gully deepens where the water strikes the lower elevation, and the ridge itself is eroded in the uphill direction. Freezing and thawing can increase erosion by loosening the sides of gullies until pieces of soil slough off and fall towards the gully bottom. Finally, the sides of gullies may fall downward due to gravitational forces when soil cohesion has been weakened by excess soil moisture.

Figure 39. Rill and gully erosion. The shallow parts of these erosion channels can be eliminated with conventional tillage and are therefore rills. The deeper parts cannot be eliminated in the course of conventional tillage and are therefore gullies.

Gully erosion is the most spectacular form of erosion and is the most difficult and expensive to control. It is wise for reclamationists to inspect all lands under their control at frequent intervals in order to detect early erosion before gully formation. Some sheet and rill erosion are inevitable, but not gully erosion.

RAINFALL

Rain and snow provide most of the water that reaches minesoils, but most erosion problems are associated with rainfall alone. Rain causes few erosion problems if the rainfall is slow enough to soak into the soil without running over the soil surface. However, most rains either fall too fast for the soil to absorb all the water or the soil becomes saturated and subsequent rain is unable to soak into the soil. In either case the excess water flows downhill over the soil surface and probably causes some erosion.

The rainfall characteristics that have the most effect on erosion are: (1) intensity (amount of rain that falls within a given time); (2) duration (length of time that rainstorm occurs); (3) intensity of rainfall during different periods of the rainstorm; (4) frequency of rainstorm occurrence; and (5) amount of rainfall during each part of the year.

1. Rainfall intensity is expressed as inches per hour (in./h). If 1/2 inch of rain fell in 15 minutes, the intensity would be 2 in./h.

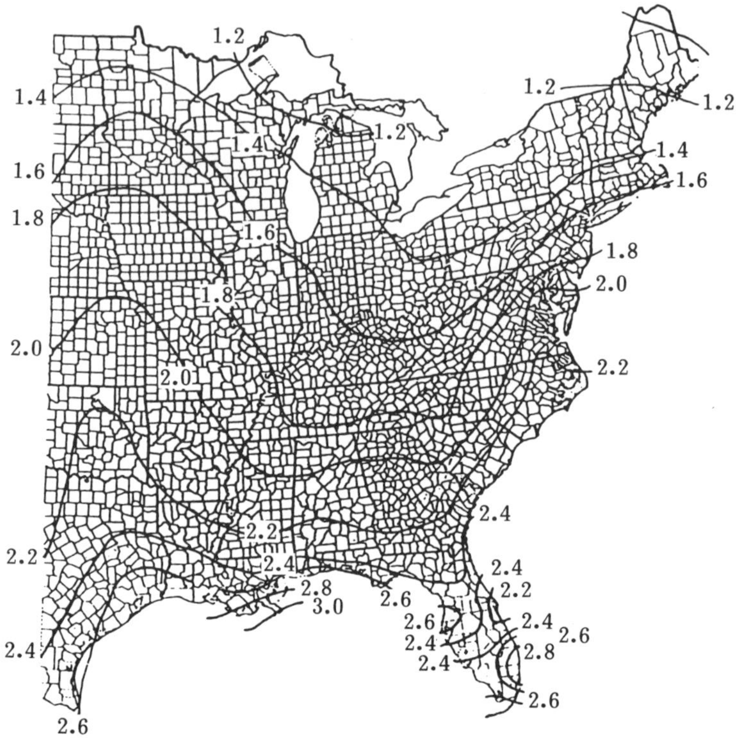

Figure 40. Maximum amount of rainfall, in inches, expected in a 10-year period for a 30-minute duration storm. *Source:* Hershfield, D.M. 1961. *Rainfall Frequency Atlas of the United States.* U.S. Weather Bureau Tech. Paper 40.

If 2 inches fell in 3 hours, the intensity would be 2/3 in./h. The more intense the rainfall (the greater the amount of rain per hour), the more damage the rainfall can do for a given period of time. A less intense rain over a much longer period of time may do more damage, but for any given period of time the more intense rain will do more damage.

The maximum rainfalls expected for given periods of time have been calculated for various parts of the U.S. Figure 40 is a map showing the maximum inches of rainfall that can be expected from a storm lasting 30 minutes in any 10-year period. The line passing through the northern part of Alabama shows that the most rain expected from a rainstorm lasting 30 minutes in any 10-year interval would be 2 inches. This would be an intensity of 4 in./h. Figure 41 shows the maximum rainfall expected for a storm lasting 24 hours during a 10-year period in the eastern U.S. This is the period used by most surface mine regulatory authorities to plan erosion and sedimentation control.

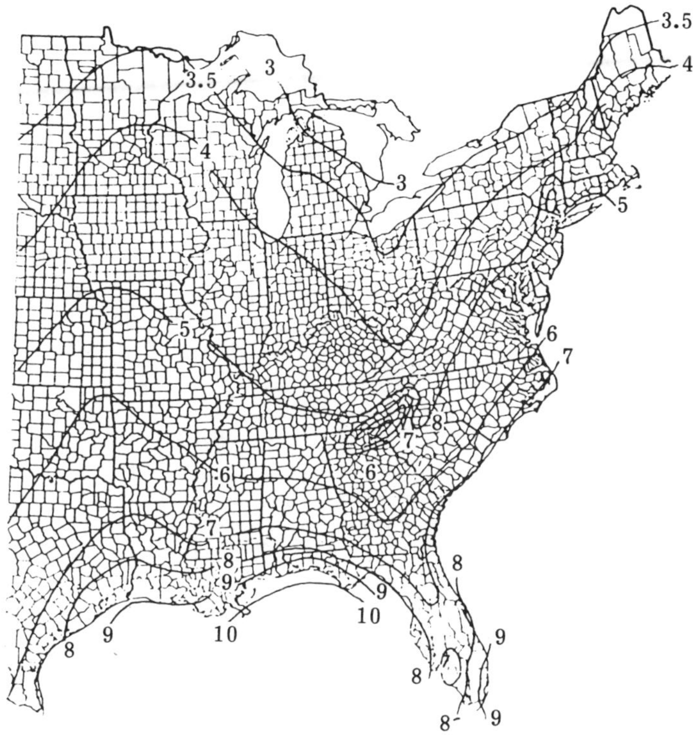

Figure 41. Maximum amount of rainfall, in inches, expected in a 10-year period for a 24-hour duration storm. *Source:* Hershfield, D.M. 1961. *Rainfall Frequency Atlas of the United States.* U.S. Weather Bureau Tech. Paper 40.

2. The duration of high-intensity rainfalls is usually shorter than the duration of low-intensity rainfalls. It is well known that hard, fast rainfalls do not last as long as soft, slow rainfalls.

3. The intensity of a rainstorm during different parts of the storm is important because it has an effect on the amount of water that soil can absorb. If the storm is intense at the beginning, the large soil pores are likely to be empty and can absorb the rain at a maximum rate. If the greatest intensity occurs at the end of the storm, the soil pores will already be filled with water and infiltration will be slow or completely stopped. Water from this intense rainfall will then flow over the soil surface, causing erosion.

4. The frequency of rainstorms is also important because of the effect on water infiltration. When rains occur frequently, there is not enough time for the water from previous rains to move downward into deeper storage horizons or the water table. When the

soil pores are saturated, the surplus water from succeeding rainfalls will flow downhill over the soil surface, causing erosion.

5. It is best for the reclamationist if the more intense rainfalls occur after a vegetation cover has been established. If intense rains occur before seedling establishment, there is likely to be a maximum amount of soil erosion and soil water loss.

RUNOFF

Runoff is that part of the rainfall that does not infiltrate the soil and flows over the soil surface to a channel and on into a sediment pond or other body of water. The more easily a rain enters the soil surface and the more storage space (pores) in the soil, the less water available for runoff. Sandy soils have large pores and usually have a higher infiltration rate than silty or clayey soils. If the sandy topsoil has a compacted subsoil of overburden material, the rainfall will probably be absorbed quickly until the sandy topsoil becomes saturated, at which time runoff will occur. If the overburden beneath the topsoil is not compacted, the rain will move into the overburden after filling the topsoil, and much more rain will be absorbed before runoff can occur.

Any management practice that improves soil structure also results in spaces within the soil for rain infiltration. The resulting water storage will decrease the amount of runoff. A decrease in the runoff means less soil erosion and more water stored in the soil for seed germination and plant growth.

Cultivation can be used to loosen surface soil that has been compacted and increase the rate of rain infiltration. However, this improvement is only temporary if there is no live vegetation or mulch on the soil to prevent the rain from recompacting the surface soil. Subsoiling can be used to loosen the soil to a greater depth than that obtained with surface tillage. Subsoiling can be used after vegetation establishment in order to loosen subsoil or overburden layers that are restricting rain infiltration. All tillage should be done on the contour in order to have furrows oriented in such a way that they will intercept water flowing downhill. The ridge caused by plowing will slow the water and provide more time for the water to infiltrate the soil. Furrows running uphill and downhill will provide beginning channels for erosion gullies and should be avoided.

Vegetation can significantly increase infiltration. Rain falling on a vegetated soil does not pack the soil and does not fill the soil pores with fine soil material. A vegetated soil tends to retain its permeability throughout a rainstorm until the pores are filled with water. Runoff does not occur on a vegetated soil until the soil pores become saturated with water. When runoff does occur on a vegetated soil, the vegetation at the soil surface slows the runoff and traps soil particles that might otherwise move downhill with the water. However, this does not help the reclamationist through that critical time after the topsoil has been

Table 18. Expected Peak Rates of Runoff from Watersheds[1]

Acres	Runoff[2]	Acres	Runoff	Acres	Runoff
1	4	75	195	260	510
2	8	80	205	280	540
3	12	85	215	300	570
4	15	90	225	320	598
5	18	95	235	340	626
10	36	100	245	360	651
15	52	110	265	380	682
20	67	120	285	400	710
25	81	130	304	420	736
30	94	140	322	440	762
35	107	150	340	460	788
40	120	160	356	480	814
45	132	170	372	500	840
50	144	180	388	520	864
55	155	190	404	540	888
60	166	200	420	560	912
65	175	220	450	580	939
70	185	240	480	600	960

[1]The following conditions apply: 10-year frequency runoff calculated for areas located along line 1.0, Figure 42; soils with an average infiltration rate; average land slope, 8 percent; typical shape; row crops planted across the slope without terraces; no appreciable surface storage.

[2]Runoff data given as cubic feet per second (cfs).

Source: Beasley, R.P. 1972. Erosion and Sediment Pollution Control. Ames, IA: Iowa State University Press. Used by permission.

replaced and before vegetation has become established. This critical period must be endured, but there are precautions that the reclamationist can take to prevent or reduce the amount of soil erosion.

The reclamationist needs to know the greatest amount of water that can be expected to concentrate at the outlet of a watershed during a rainstorm. This information is needed to ensure that the erosion control structures have been designed to cope with this amount of water. A watershed is an area that delivers all its runoff to a single exit point. A watershed can also be described as the total land area above a given point that contributes runoff to that point. Several small watersheds may be considered one larger watershed if they all eventually empty their runoff at one point.

Erosion control structures such as diversion ditches are usually designed to carry the greatest amount of runoff that is expected within a specified period, eg, 5, 10, 25, 50, or 100 years. The most accurate method of predicting runoff from an area is to know the past runoff history

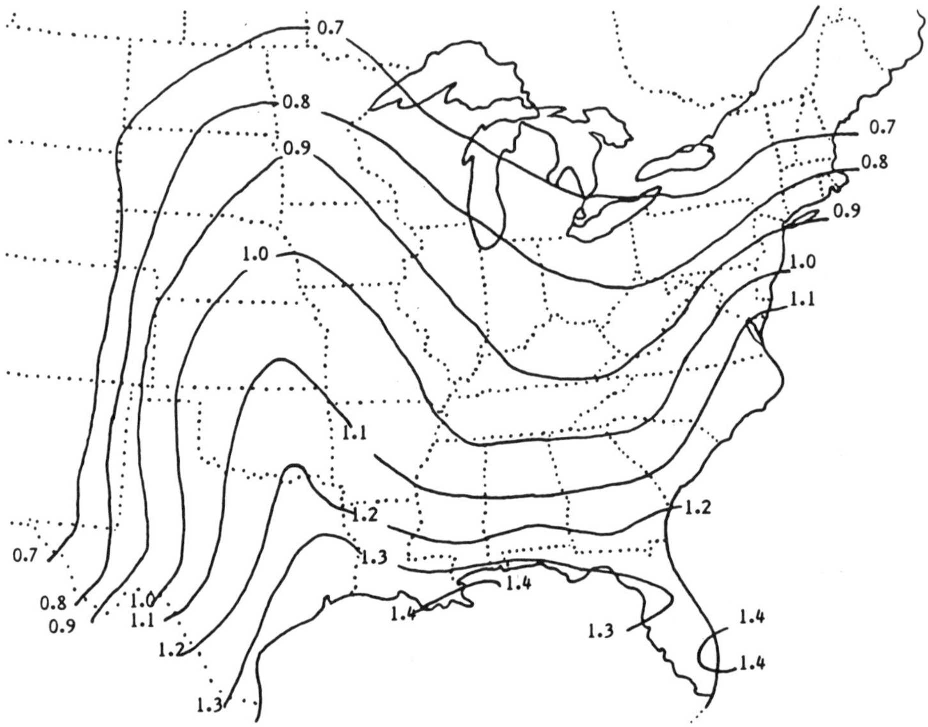

Figure 42. Values for the location factor (L). *Source:* Beasley, R.P. 1972. *Erosion and Sediment Pollution Control.* Ames, IA: Iowa State University Press. Used by permission.

for the area. Of course, this is not possible for surface mine reclamation because the watersheds are newly formed in the process of reclamation.

R.P. Beasley (1972) developed an equation that can be used to estimate the peak runoff that can be expected from a watershed. The equation is:

$$Q = Q_T \times L \times I \times T \times S \times V \times C \times P \times F$$

where:

Q = peak rate of runoff in cubic feet per second (cfs)

Q_T = peak rate of runoff from a hypothetical watershed under a specific set of conditions

L = watershed location

I = soil infiltration

T = topography

Table 19. Topographic Factor (T)

Average land slope (percent)	T
1	0.65
2	0.72
3	0.78
4	0.83
5	0.88
6	0.92
7	0.96
8	1.00
9	1.04
10	1.07
12	1.14
14	1.20
16	1.26
18	1.32
20	1.37
22	1.43
24	1.49
26	1.54
28	1.59
30	1.63
32	1.67
34	1.71

Source: Beasley, R.P. 1972. Erosion and Sediment Pollution Control. Ames, IA: Iowa State University Press. Used by permission.

S = watershed shape

V = vegetative cover

C = contour farming

P = surface storage

F = runoff frequency

The values to be used for Q_T are shown in Table 18. These peak rate runoff values have been calculated for a 10-year frequency for the areas located along line 1.0 on the map shown in Figure 42. These calculations are based on soils with an average infiltration rate, average land slope (8 percent), and typical watershed shape with a row crop planted across the slope but no terraces, diversions, or appreciable surface storage.

Table 20. Watershed Shape Factor (S)

Acres	Maximum distance traveled by runoff (ft)										
5				400	500	600	700	800	900	1,100	1,300
10		500	550	600	700	800	950	1,100	1,300	1,600	1,900
20	600	700	800	900	1,000	1,200	1,400	1,600	1,900	2,300	2,800
30	800	900	1,000	1,200	1,400	1,600	1,900	2,200	2,600	3,100	3,600
40	1,000	1,100	1,200	1,400	1,600	1,900	2,200	2,600	3,100	3,600	4,200
50	1,100	1,200	1,400	1,600	1,800	2,100	2,400	2,900	3,100	3,900	4,600
60	1,300	1,400	1,600	1,800	2,100	2,400	2,700	3,200	3,700	4,200	5,000
70	1,400	1,600	1,800	2,000	2,300	2,700	3,100	3,600	4,100	4,600	5,400
80	1,500	1,700	1,900	2,100	2,500	2,900	3,300	3,800	4,400	5,000	5,800
90	1,700	1,800	2,000	2,300	2,700	3,100	3,600	4,100	4,600	5,400	6,200
100	1,800	1,900	2,200	2,500	2,900	3,300	3,800	4,400	5,000	5,700	6,600
120	1,900	2,100	2,400	2,800	3,200	3,600	4,100	4,700	5,400	6,200	7,100
140	2,100	2,400	2,700	3,100	3,600	4,000	4,500	5,100	5,900	6,700	7,600
160	2,300	2,600	3,000	3,400	3,800	4,300	4,800	5,500	6,300	7,100	8,000
180	2,500	2,900	3,300	3,700	4,100	4,600	5,200	5,900	6,700	7,500	8,400

200	2,700	3,100	3,500	3,900	4,500	4,800	5,500	6,200	7,000	7,900	8,800
250	3,200	3,600	4,000	4,400	4,900	5,500	6,300	7,100	7,900	8,700	9,600
300	3,700	4,100	4,500	5,000	5,500	6,200	7,000	7,700	8,500	9,400	10,400
350	4,100	4,500	4,900	5,400	6,100	6,800	7,500	8,200	9,000	10,000	11,200
400	4,500	4,900	5,400	6,000	6,700	7,400	8,200	9,000	9,900	10,800	12,000
450	4,900	5,400	5,900	6,500	7,200	7,900	8,700	9,600	10,600	11,600	12,800
500	5,400	5,900	6,500	7,100	7,700	8,400	9a,200	10,100	11,100	12,200	13,600
550	5,900	6,400	7,000	7,600	8,800	9,000	9,800	10,700	11,600	12,900	14,500
600	6,300	6,900	7,500	8,100	8,800	9,600	10,400	11,300	12,400	13,700	15,100
Shape factor S											
	1.25	1.20	1.15	1.10	1.05	1.00	0.95	0.90	0.85	0.80	0.75

Source: Beasley, R.P. 1972. Erosion and Sediment Pollution Control. Ames, IA: Iowa State University Press. Used by permission.

The values to be used for *L* are shown in Figure 42. This factor is based on the rainfall intensity that can be expected in various parts of the U.S. Rainfall intensity has an appreciable effect on rate of runoff.

The values to be used for *I* are determined by the texture of the topsoil (surface soil) and subsoil (regraded overburden).

1. Infiltration very high—coarse-textured soils in topsoil and subsoil (sands and loamy sands); $I = 0.8$
2. Infiltration above average—topsoil and subsoil medium to moderately coarse-textured (sandy loam); $I = 0.9$
3. Infiltration average—medium-textured topsoil (loam, silt loam, silt), moderately fine-textured subsoil (silty clay loam, clay loam, sandy clay loam); $I = 1.0$
4. Infiltration below average—medium-textured (silt, silt loam, loam) to moderately fine-textured (silty clay loam, sandy clay loam, clay loam) topsoil with fine textured subsoil (silty clay, clay, sandy clay); $I = 1.1$
5. Infiltration very low—topsoil and subsoil fine-textured (silty clay, clay, sandy clay) or compacted; $I = 1.2$

The values to be used for *T* depend on the steepness of the slope. Runoff has less opportunity to infiltrate the soil surface as the slope becomes more steep due to the increase in velocity of runoff. Table 19 gives the values of *T* to be used in the equation.

The values to be used for watershed shape (*S*) are given in Table 20. These values depend on the size of the watershed and the maximum distance that the runoff must flow to leave the watershed. Distance traveled in diversion ditches should be included in the distance value. The length of the diversion ditch is added to the greatest distance that water will have to flow to reach the upper end of the ditch. Values for *S* are shown at the bottom of the table.

The values used for *V* depend on the amount of vegetative cover on the minesoil. The actual value used is chosen from the following list: urban areas, $V = 2.0$; farmsteads and suburbs, $V = 1.2$; small grain of good quality, $V = 0.8$; small grain of poor quality, $V = 0.9$; row crops, $V = 1.0$; pasture of good quality, $V = 0.6$; pasture of poor quality, $V = 0.8$; meadow of good quality, $V = 0.5$; meadow of poor quality, $V = 0.7$; timber of good quality, $V = 0.4$; timber of poor quality, $V = 0.6$. When the equation is used by reclamationists to calculate the peak rate of runoff for bare minesoil, the value $V = 1.5$ should probably be used because the values in Table 18 are based on a soil with crop rows across the slope. This would provide less runoff than a bare minesoil.

The values for C are determined by the amount of farming that is done on the contour in the watershed. The following C values apply if the crop rows are parallel to contour lines, the grade in the crop rows

Table 21. Storage Factor for Graded Terraces (P)

Acres in watershed	Average length of terraces (ft) 500	1000	1500
0–10	0.95	0.90	0.80
11–40	0.97	0.93	0.85
41–100	0.98	0.95	0.90
101–200	0.99	0.97	0.95
201–400	1.00	0.98	0.96
401–600	1.00	0.99	0.97

Source: Beasley, R.P. 1972. Erosion and Sediment Pollution Control. Ames, IA: Iowa State University Press. Used by permission.

does not exceed 2 percent, and the crop rows discharge into grassed waterways. If the watershed is 0 to 10 acres, the C value = 0.95; 11 to 40 acres, C = 0.96; 41 to 100 acres, C = 0.97; 101 to 200 acres, C = 0.98; 201 to 600 acres, C = 0.99. If all of the watershed is not farmed on the contour, move to the next higher C value. For instance, if the watershed contains 150 acres but only 100 acres are farmed on the contour, the C value would be 0.99 instead of 0.98. In most cases, the C factor will not apply to surface mine reclamation and a value of 1.0 should be used.

The surface storage factor (P) is used to account for any runoff that is detained or impounded by terraces, ponds, or diversions. The most likely structure that the reclamationist will be using is the diversion ditch. Table 21 shows the values of P to be used for watersheds of various acreages and diversions or terraces of varying lengths. If the length is less than 500 feet, use a value of 1.00.

If a pond or other impoundment occurs in the watershed, the peak rate of runoff can be determined by:

1. Determining the peak rate of runoff expected from the area below the impoundment
2. Determining the peak rate of runoff expected from the spillway of the impoundment
3. Adding these two runoffs together to determine peak rate of runoff for the entire watershed

The value of runoff frequency (F) is used to account for a frequency different from the frequency used to obtain the Q_T value. The Q_T value is based on a 10-year frequency. If the period to be used is 0.5 year, F = 0.2; 1 year, F = 0.3; 2 years, F = 0.5; 5 years, F = 0.8; 10 years, F = 1.0; 25 years, F = 1.2; 50 years, F = 1.5.

Table 22. Rational Runoff Coefficients

Topography and vegetation	Value of C in $Q = CIA$[1]		
	Soil texture		
	Open sandy loam	Clay and silt loam	Tight clay
Woodland			
Flat (0-5)[2]	0.10	0.30	0.40
Rolling (5-10)	0.25	0.35	0.50
Hilly (10-30)[3]	0.30	0.50	0.60
Pasture			
Flat	0.10	0.30	0.40
Rolling	0.16	0.36	0.55
Hilly	0.22	0.42	0.60
Cultivated			
Flat	0.30	0.50	0.60
Rolling	0.40	0.60	0.70
Hilly	0.52	0.72	0.82

[1]C = constant; Q = peak rate of runoff in cubic feet per second; I = rainfall intensity for time of concentration in inches per hour; A = area of watershed in acres. See text for further explanation.

[2]Numbers in parentheses indicate percent of slope.

[3]Values are not available for steeper slope conditions. Therefore, applying the rational formula to steeper slopes, this limitation must be realized.

Source: Frevert, R.K. et al., 1955. Soil and Water Conservation Engineering. New York: Wiley.

Another method of determining the peak rate of runoff is by applying the rational formula.

$$Q = CIA$$

Q = peak rate of runoff in cubic feet per second (cfs)

C = constant (Table 22)

I = rainfall intensity for the time of concentration in inches per hour (in./h)

A = area of watershed in acres

In this equation the constant value C is used to account for the soil texture that affects the ease with which water can enter the soil and the amount of water that can be stored without runoff. This constant also

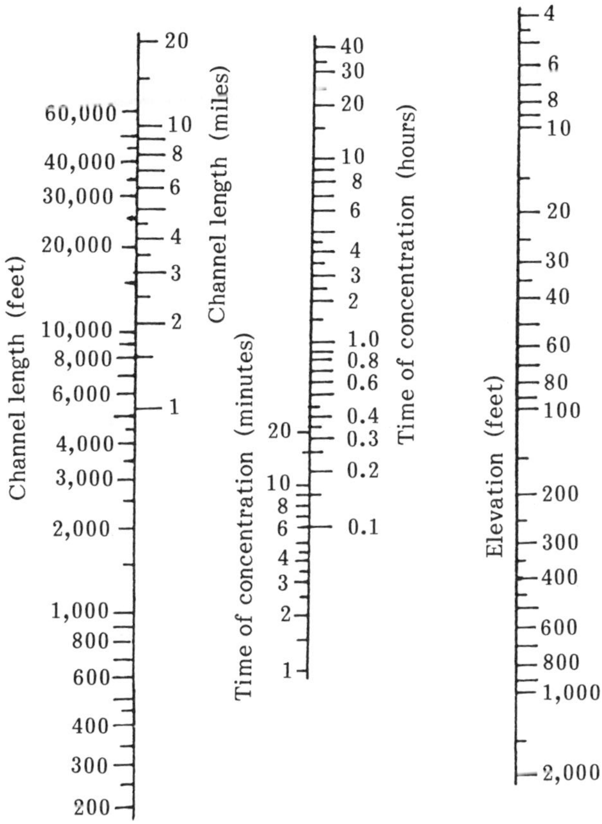

Figure 43. Nomograph for estimating time of concentration. Modified from Kirpich, P.Z. 1940. *Civil Engineering,* 10:6.

accounts for the topography and type of cover found on the watershed. As the amount of cover on a watershed increases, the amount of runoff decreases.

The value for I is the maximum rate of rainfall expected on the watershed when time for the runoff to concentrate at the watershed outlet is considered. The time of concentration is affected by the size, shape, and topography of the watershed.

The size of the watershed also has an obvious effect on the amount of water that may occur in the runoff. This is accounted for by including the watershed area (A) in the equation.

Example: A surface coal miner near Birmingham, Alabama, has a watershed of 55 acres. The length of the longest channel draining the watershed is 2200 feet, and the highest point in the watershed is 200 feet above the watershed outlet. The watershed has been topsoiled and 1 ton of mulch per acre has been applied. The soils are sandy

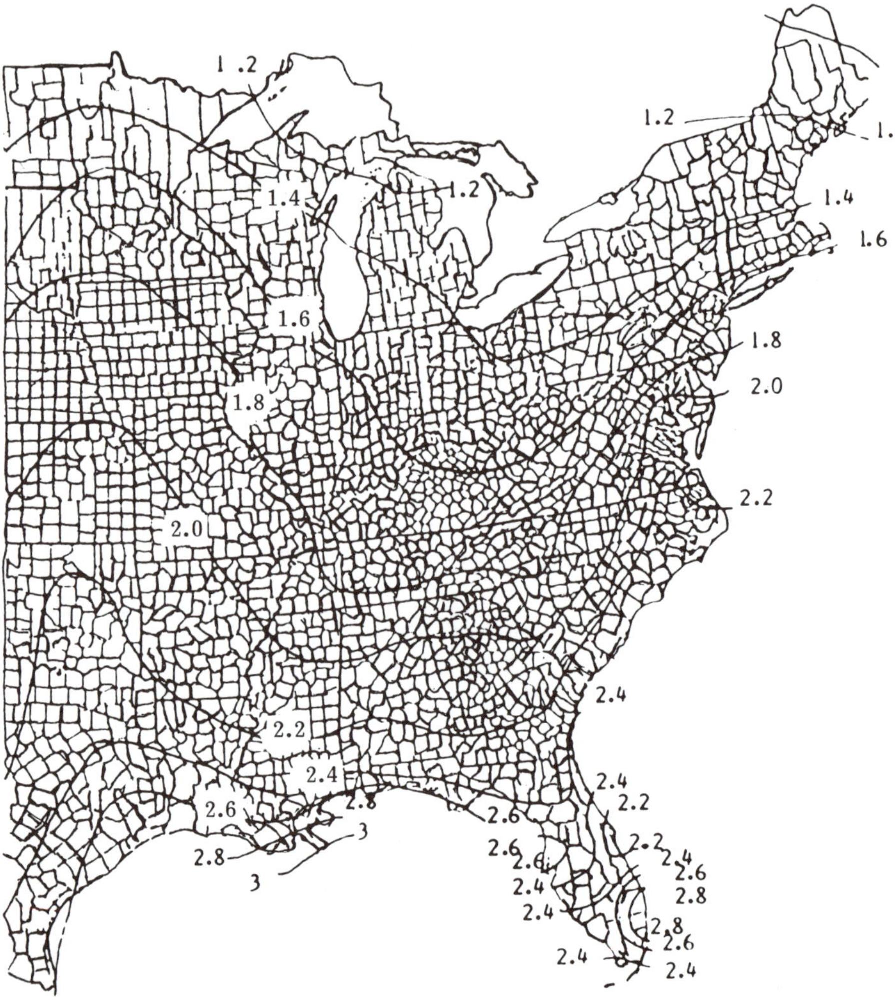

Figure 44. Maximum amount of rainfall, in inches, expected in a 10-year period for 30-minute duration rainstorm. *Source:* Hershfield, D.M. 1961. *Rainfall Frequency Atlas of the United States.* U.S. Weather Bureau Tech. paper 40.

and have a 15 percent average slope. Determine the peak rate of runoff for a 10-year frequency.

The value for *C* can be obtained from Table 22. The mulch should make the watershed comparable to a pasture and the value for a sandy pasture soil with 15 percent slope is 0.22. If the soil is compacted and the infiltration rate reduced, the *C* value should probably be increased to that of the next finer texture (0.42).

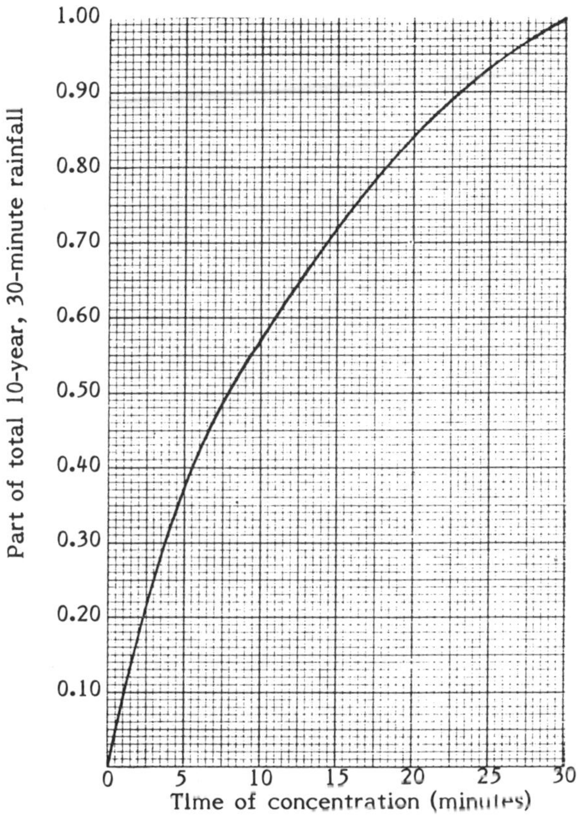

Figure 45. Parts of a 10-year, 30-minute rainfall that would fall during various times of concentration.

To obtain the rainfall intensity, the time of concentration must be determined. To do this the longest channel (2200 feet) and highest point (200 feet) values from the nomograph in Figure 43 are applied. The indicated time of concentration is 7.5 minutes. Figure 44 shows that the maximum rainfall expected from a 30-minute rainfall in a 10-year period for Birmingham is 2.00 inches. However, the value needed is the amount of rainfall for the 7.5-minute time of concentration. To obtain this value Figure 45 is used to determine the part of the 30-minute rain that would fall in a 7.5-minute rain. Figure 45 shows that the rainfall for 7.5 minutes would be 0.47 that of a 30-minute rain. Therefore, the rain that would fall in 7.5 minutes would be: 0.47 x 2.00 inches = 0.94 inches. The intensity of this rainstorm would be:

$$\frac{60 \text{ min/h}}{7.5 \text{ min}} \times 0.94 \text{ in.} = 7.52 \text{ in./h}$$

Table 23. Soil Groups Based on Water Infiltration and Transmission Characteristics for Curve Number (CN) Determination

Soil group	Soil description
A	Soils with a high infiltration rate (5 in./h or greater)—excessively drained sands or gravels. Transmission rate of water within the soil is also high. Tend to be deep sands and deep well-aggregated silt loams.
B	Soils with moderate rates of infiltration and transmission (0.2 to 5.0 in./h)—moderately deep to deep, moderately well- to well-drained soils with moderately fine to moderately coarse textures. Tend to be sandy loams and silt loams.
C	Soils with slow rates of infiltration and transmission (0.05 to 0.2 in./h)—soils with a compacted layer that impedes downward movement of water or soils with a moderately fine to fine texture. Tend to be high in clay, low in organic matter.
D	Soils with very slow rates of infiltration and transmission (less than 0.05 in./h)—heavy clay soils, soils with a high water table, highly compacted soils, and soils overlying solid rock. Tend to be very high in clay.

All of the values are now available for the determination of peak runoff for this watershed:

$$\begin{aligned} Q &= CIA \\ &= 0.22 \times 7.52 \text{ in./h} \times 55 \text{ acres} \\ &= 90.99 \text{ cfs} \end{aligned}$$

The U.S.D.A. Soil Conservation Service has developed a simple and reliable method for determining peak rates of runoff for watersheds up to 2000 acres. To use this method, the following must be known: size of the watershed, a value known as the curve number (CN), and whether the average slope is flat, moderate, or steep. A flat slope is 0 to 3 percent, a moderate slope 3 to 8 percent, and a steep slope greater than 8 percent.

The effect of soil characteristics, vegetative cover, and mechanical erosion control measures must be taken into account and this is done through the use of curve numbers (CN). Table 23 shows the soil groups to be used for CN determinations. Table 24 shows the CN to be used

Table 24. Runoff Curve Numbers (CNs) for Selected Agricultural, Suburban, and Urban Land Use

Land use description	Hydrologic soil group			
	A	B	C	D
Cultivated land[1]				
Without conservation treatment	72	81	88	91
With conservation treatment	62	71	78	81
Grasses and legumes grown for grazing				
Pasture or range land				
Poor condition	68	79	86	89
Good condition	39	61	74	80
Grasses and legumes grown for hay				
Meadow: good condition	30	58	71	78
Wood or forest land				
Thin stand, poor cover, no mulch	45	66	77	83
Good cover [2]	25	55	70	77
Open spaces, lawns, parks, golf courses, cemeteries, etc.				
Good condition: grass cover on 75 percent or more of the area	39	61	74	80
Fair condition: grass cover on 50 percent to 75 percent of the area	49	69	79	84
Commercial and business areas (85 percent impervious)	89	92	94	95
Industrial districts (72 percent impervious)	81	88	91	93
Residential:[3]				
Average lot size — Average percent impervious[4]				
1/8 acre or less — 65	77	85	90	92
1/4 acre — 38	61	75	83	87
1/3 acre — 30	57	72	81	86
1/2 acre — 25	54	70	80	85
1 acre — 20	51	68	79	84
Paved parking lots, roofs, driveways, etc.	98	98	98	98

continued

Table 24. Runoff Curve Numbers (CNs) for Selected Agricultural, Suburban, and Urban Land Use *(Continued)*

Land use description	Hydrologic soil group A	B	C	D
Streets and roads				
Paved with curbs and storm sewers[5]	98	98	98	98
Gravel	76	85	89	91
Dirt	72	82	87	89

[1]For a more detailed description of agricultural land use CNs, refer to National Engineering Handbook, Section 4, Hydrology, Chapter 9, Aug. 1972.

[2]Good cover is protected from grazing and litter and brush cover soil.

[3]Curve numbers are computed assuming the runoff from the house and driveway is directed toward the street with a minimum of roof water directed to lawns where additional infiltration could occur.

[4]The remaining pervious areas (lawn) are considered to be in good pasture condition for these CNs.

[5]In some warmer climates of the country, a CN of 95 may be used.

Source: U.S. Department of Agriculture, Soil Conservation Service. 1975. Urban Hydrology for Small Watersheds, Technical Release No. 55. Washington, D.C.: U.S. Government Printing Office.

based on soil groups and the land use or condition. Most minesoils will be classed as cultivated without conservation measures unless diversion ditches or some other mechanical erosion control methods are used. In the following example the minesoil will be classified as cultivated with conservation treatment.

The CN determined from Table 24 can now be used with Figures 46, 47, or 48 to obtain peak runoff for the watershed. However, these peak rates of runoff are given in cubic feet per second per inch of runoff. Therefore the inches of runoff for different rainfalls must be determined. These runoff values can be obtained from Table 25.

Example: Determine the peak runoff expected from a 25 acre watershed of sandy clay loam with an infiltration rate of 1.0 in./h. The topsoil has been replaced and diversion ditches constructed. The slopes of the watershed average approximately 6 percent. The watershed is located in east Tennessee and the peak runoff is needed for a 10-year, 24-hour rainfall. Table 23 indicates that a soil with an infiltration rate of 1.0 in./h belongs in group B. Table 24 indicates that a soil

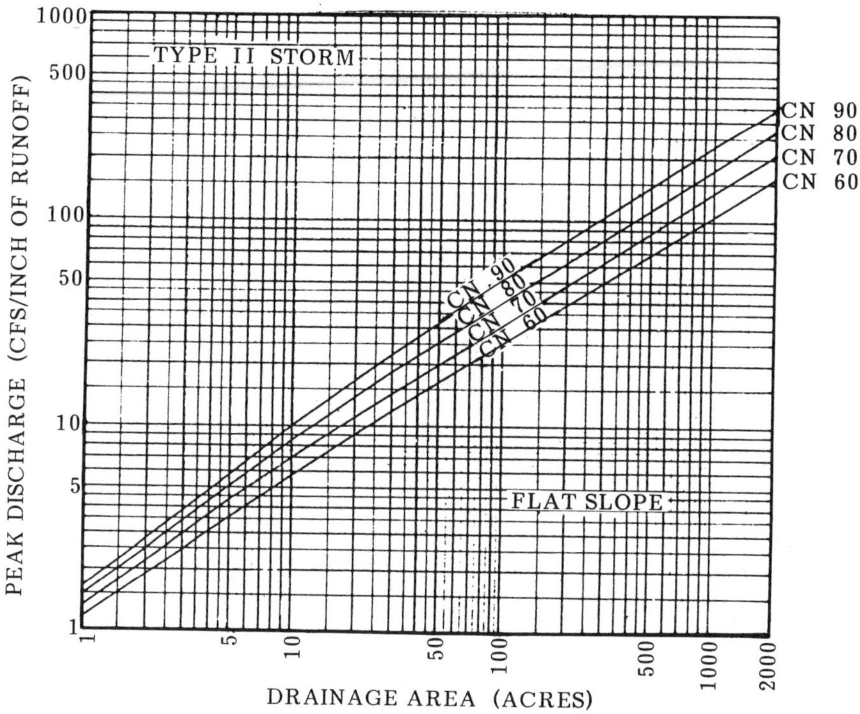

Figure 46. Peak rates of discharge for small watersheds (24-hour, type-II storm distribution). *Source:* U.S. Department of Agriculture, Soil Conservation Service. 1975. *Urban Hydrology for Small Watersheds,* Technical Release No. 55. Washington, D.C.: U.S. Government Printing Office.

from group B that is bare but has diversion ditches would have a CN of 71. Figure 47 (for moderate slopes) shows that a watershed of 25 acres with a CN of 71 would have an approximate peak runoff of 21 cfs/in. of runoff. The rainfall for a 10-year, 24-hour event in east Tennessee will be approximately 5 inches (Figure 41). Table 25 shows that a 5-inch rainfall would produce a runoff depth of 2.12 inches (interpolated) for a CN of 71. Therefore, the peak rate of runoff will be 2.12 in. x 21 cfs/in. = 44.52 cfs.

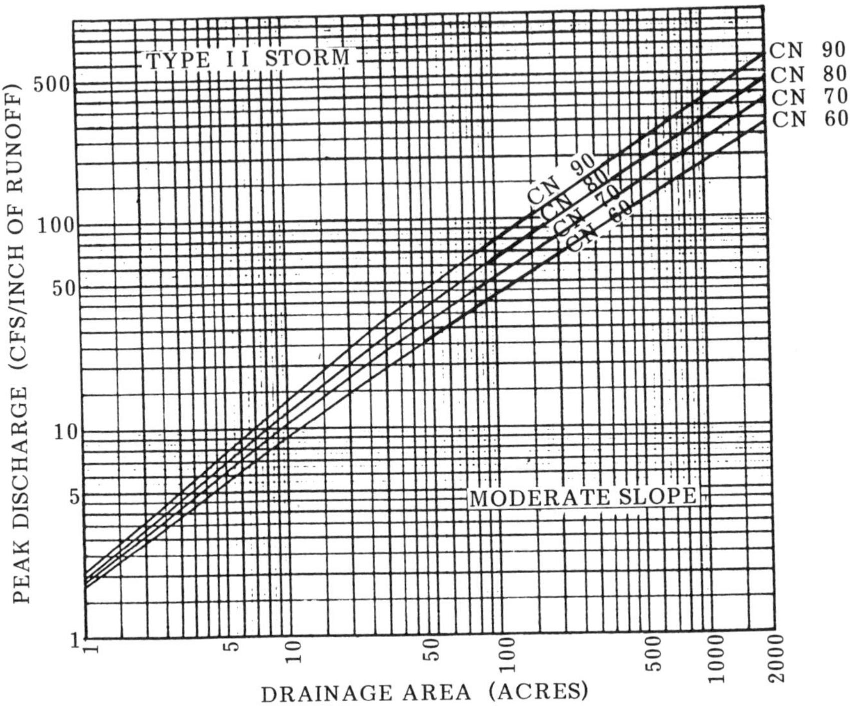

Figure 47. Peak rates of discharge for small watersheds (24-hour, type-II storm distribution). *Source:* U.S. Department of Agriculture, Soil Conservation Service. 1975. *Urban Hydrology for Small Watersheds,* Technical Release No. 55. Washington, DC: U.S. Government Printing Office.

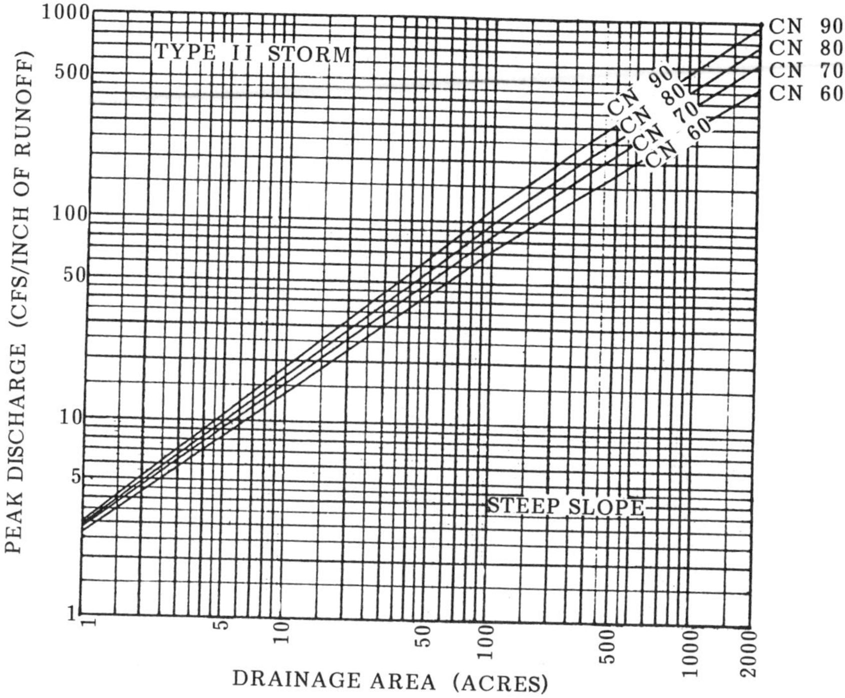

Figure 48. Peak rates of discharge for small watersheds (24-hour, type-II storm distribution). *Source:* U.S. Department of Agriculture, Soil Conservation Service. 1975. *Urban Hydrology for Small Watersheds,* Technical Release No. 55. Washington, DC: U.S. Government Printing Office.

Table 25. Runoff Depth in Inches for Selected Curve Numbers (CNs) and Rainfall Amounts

Rainfall (in.)	CN[1]								
	60	65	70	75	80	85	90	95	98
1.0	0	0	0	0.03	0.08	0.17	0.32	.56	.79
1.2	0	0	0.03	0.07	0.15	0.28	0.46	.74	.99
1.4	0	0.02	0.06	0.13	0.24	0.39	0.61	.92	1.18
1.6	0.01	0.05	0.11	0.20	0.34	0.52	0.76	1.11	1.38
1.8	0.03	0.09	0.17	0.29	0.44	0.65	0.93	1.29	1.58
2.0	0.06	0.14	0.24	0.38	0.56	0.80	1.09	1.48	1.77
2.5	0.17	0.30	0.46	0.65	0.89	1.18	1.53	1.96	2.27
3.0	0.33	0.51	0.72	0.96	1.25	1.59	1.98	2.45	2.78
4.0	0.76	1.03	1.33	1.67	2.04	2.46	2.92	3.43	3.77
5.0	1.30	1.65	2.04	2.45	2.89	3.37	3.88	4.42	4.76
6.0	1.92	2.35	2.80	3.28	3.78	4.31	4.85	3.41	5.76
7.0	2.60	3.10	3.62	4.15	4.69	5.26	5.82	6.41	6.76
8.0	3.33	3.90	4.47	5.04	5.62	6.22	6.81	7.40	7.76
9.0	4.10	4.72	5.34	5.95	6.57	7.19	7.79	8.40	8.76
10.0	4.90	5.57	6.23	6.88	7.52	8.16	8.78	9.40	9.76
11.0	5.72	6.44	7.13	7.82	8.48	9.14	9.77	10.39	10.76
12.0	6.56	7.32	8.05	8.76	9.45	10.12	10.76	11.39	11.76

[1]To obtain runoff depths for CNs and other rainfall amounts not shown in this table, use an arithmetic interpolation.

Source: U.S. Department of Agriculture, Soil Conservation Service. 1975. Urban Hydrology for Small Watersheds, Technical Release No. 55. Washington, D.C.: U.S. Government Printing Office.

CHAPTER 14

SOIL EROSION PREDICTION

An equation termed the universal soil loss equation has been developed that can be used to make a reasonable estimate of the amount of sheet and rill erosion to be expected on a surface mine. This equation can also be used to determine actions to be taken in the process of reclamation. For instance, the need for diversion ditches can be determined by this equation and the length and steepness of slopes that should be created in the regrading process. The value obtained with this equation is *not* the total erosion that takes place on the slope under study. This equation estimates only the amount of soil that will be removed from the slope by the erosion process. Part of the soil that erodes on the slope will move from its original position but will not actually leave the slope. The slope factor in the equation describes the location from which soil will be totally removed. This equation has been presented in several forms and the following form appears to be the one most useful for surface mine reclamation:

$$E = RKLSCP$$

E = amount of erosion in tons per acre per year (ton/A/yr)

R = rainfall factor

K = soil erodibility factor

L = slope length factor

S = slope gradient factor

C = vegetation factor

P = mechanical erosion control factor

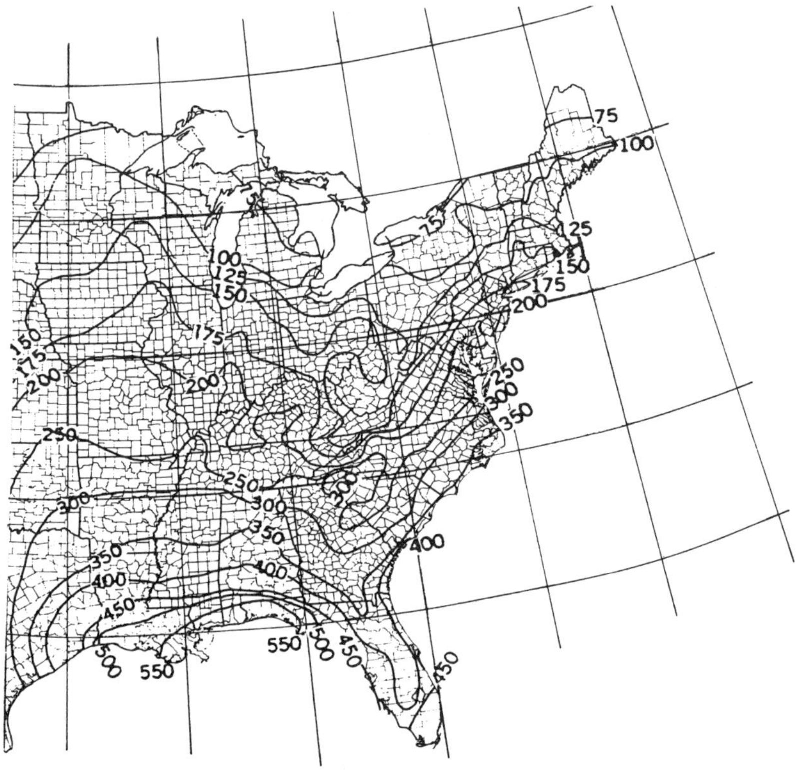

Figure 49. Average annual values of the rainfall erosion index (R). *Source:* Wischmeir, W.H. and D.D. Smith. 1978. *Predicting Rainfall Erosion Losses.* U.S.D.A. Agr. Handbook 537.

RAINFALL FACTOR

Values for the rainfall factor (R) can be obtained from Figure 49. The lines on this map are called isoerodents and they show the regions that have approximately equal erosive rainfalls. The R value for locations between isoerodents can be obtained by interpolation. The numerical value of R is an expression of the erosive effect of raindrop impact, and amount and rate of runoff of all the storms that occur in a given period of time.

SOIL ERODIBILITY FACTOR

The soil erodibility factor (K) refers to the ease with which a soil will erode. The ease with which a soil will erode is controlled by inherent characteristics of a soil such as texture, structure, permeability, and organic matter content. The value of K can be determined by applying the nomograph in Figure 50. This would require that the percents of

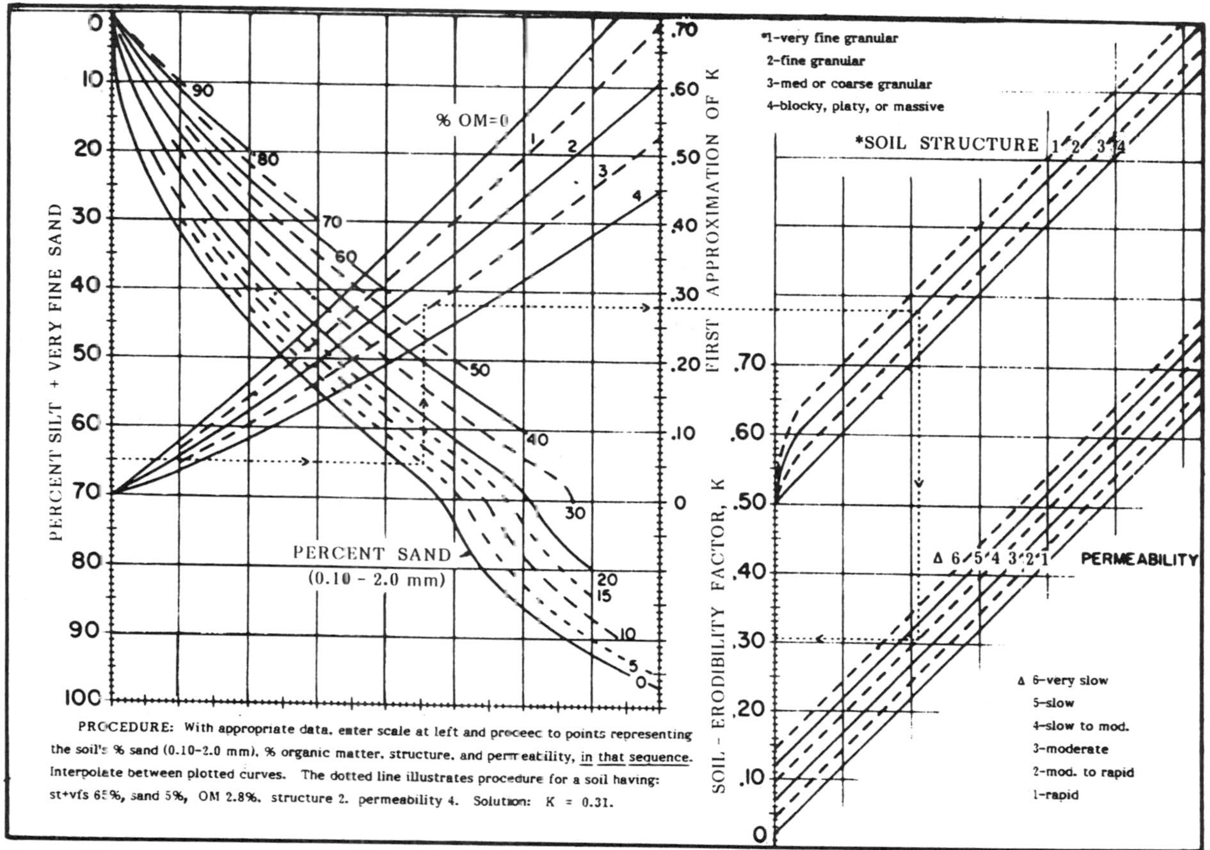

Figure 50. Soil erodibility (K) nomograph. *Source:* Wischmeir, W.H., and Smith, D.D. 1978. *Predicting Rainfall Erosion Losses.* U.S.D.A. Agr. Handbook 537.

silt, very fine sand, total sand, and organic matter be determined as well as the soil structure and permeability. These values can be determined by laboratory procedures or they can be estimated. When moist or wet soils are rubbed between the thumb and forefinger, the approximate amounts of sand and silt can be determined by feel. This method takes practice but can be learned with the help of a soil classifier from the Soil Conservation Service.

Four structure classes are used in the nomograph. Again, soil classifiers from the Soil Conservation Service can teach the reclamationist to identify the different soil structures. These classes and their definitions are:

1. Very fine granular. Soil is made up of many small clods of soil that are less than 1 mm in diameter. These clods are irregular in shape and tend to be solid rather than porous.
2. Fine granular. Same as very fine granular except 1 to 2 mm in diameter.
3. Medium or coarse granular. Same as very fine granular except 2 to 10 mm in diameter.
4. Blocky, platy, or massive. Blocky soils are made of many clods that appear to have flat or approximately flat faces covering the clod surfaces. The clods tend to be approximately the same length on all faces and can be from less than 5 mm to more than 50 mm in diameter. Platy soils are composed of plate-like clods that are much smaller in one dimension than in the other two dimensions. The plates can be from 1 to 10 mm in thickness. Massive soils are actually soils such as compacted clays that have no structure and no clods that can be described as having uniform shapes.

Relative soil permeabilities can be measured by driving a metal cylinder into a water-saturated soil and then maintaining a 0.5-inch head of water over the soil surface within the cylinder. Water is poured onto the soil surface and the rate at which water enters the soil surface is measured in inches per hour. Do not let the water fall directly on the soil surface as this may compact the soil surface and interfere with infiltration.

Permeability Class	Infiltration Rate (in./h)
1. Rapid	5.00 to 10.00
2. Moderate to rapid	2.50 to 5.0
3. Moderate	0.80 to 2.50
4. Slow to moderate	0.20 to 0.80
5. Slow	0.05 to 0.20
6. Very slow	Less than 0.05

Table 26. Values of the Topographic Factor (LS) for Specific Combinations of Slope Length and Steepness

Percent slope	Slope length (ft)											
	25	50	75	100	150	200	300	400	500	600	800	1,000
0.2	0.060	0.069	0.075	0.080	0.086	0.092	0.099	0.105	0.110	0.114	0.121	0.126
0.5	0.073	0.083	0.090	0.096	0.104	0.110	0.119	0.126	0.132	0.137	0.145	0.152
0.8	0.085	0.098	0.107	0.113	0.123	0.130	0.141	0.149	0.156	0.162	0.171	0.179
2.0	0.133	0.163	0.185	0.201	0.227	0.248	0.280	0.305	0.326	0.344	0.376	0.402
3.0	0.190	0.233	0.264	0.287	0.325	0.354	0.400	0.437	0.466	0.492	0.536	0.573
4.0	0.230	0.303	0.357	0.400	0.471	0.528	0.621	0.697	0.762	0.820	0.920	1.01
5.0	0.263	0.379	0.464	0.536	0.656	0.758	0.928	1.07	1.20	1.31	1.52	1.69
6.0	0.335	0.476	0.583	0.673	0.824	0.952	1.17	1.35	1.50	1.65	1.90	2.13
8.0	0.475	0.701	0.859	0.992	1.21	1.41	1.72	1.98	2.22	2.43	2.81	3.14
10	0.685	0.968	1.19	1.37	1.68	1.94	2.37	2.74	3.06	3.36	3.87	4.33
12	0.903	1.28	1.56	1.80	2.21	2.55	3.13	3.61	4.04	4.42	5.11	5.71
14	1.15	1.62	1.99	2.30	2.81	3.25	3.98	4.59	5.13	5.62	6.49	7.26
16	1.42	2.01	2.46	2.84	3.48	4.01	4.92	5.68	6.35	6.95	8.03	8.98
18	1.72	2.43	2.92	3.43	4.21	3.86	5.95	6.87	7.68	8.41	9.71	10.9
20	2.04	2.88	3.53	4.03	5.00	5.77	7.07	8.16	9.12	10.0	11.5	12.9

Source: Wischmeir, W.H., and Smith, D.D. 1978. Predicting Rainfall Erosion Losses: A Guide to Conservation Planning. Agriculture Handbook No. 537, U.S. Dept. of Agriculture, Science and Education Administration. Washington, D.C.: U.S. Government Printing Office.

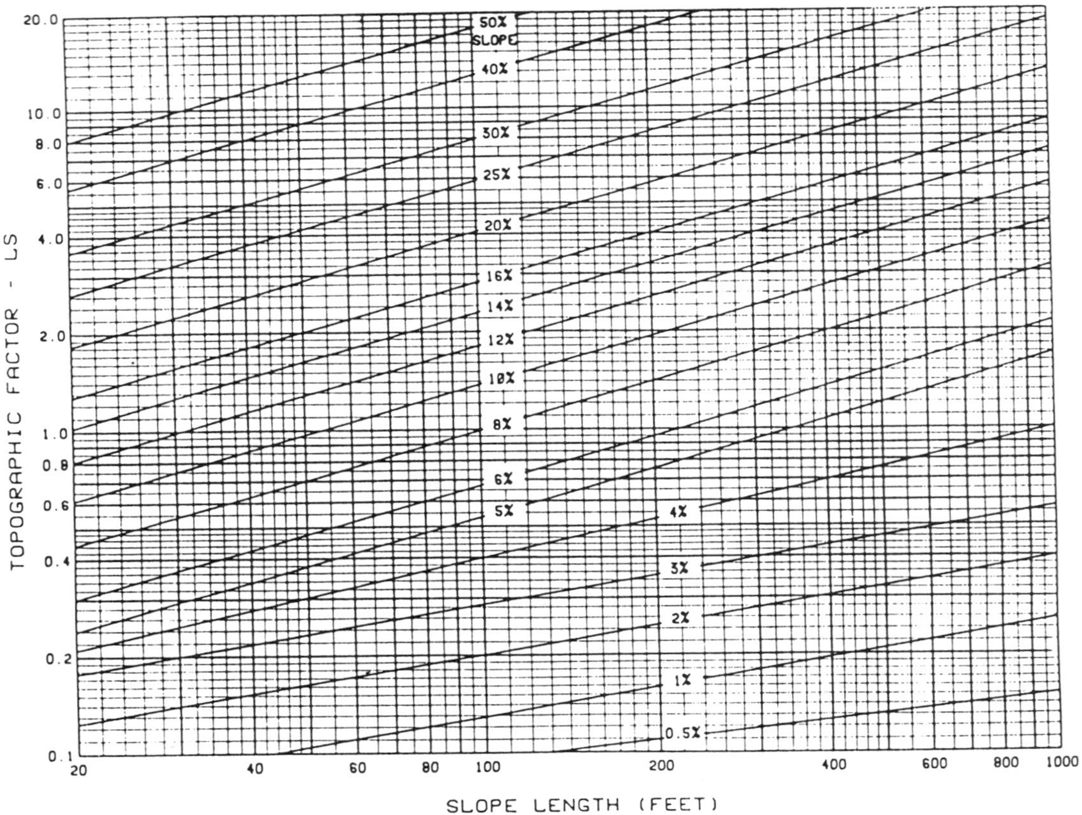

Figure 51. The effect of slope length and steepness on the topographic factor (LS). *Source:* Wischmeir, W.H., and Smith, D.D. 1978. *Predicting Rainfall Erosion Losses,* U.S.D.A. Agr. Handbook 537.

Permeability can be estimated by the reclamationist after some experience with local soils. Remember that permeability is controlled by soil texture, structure, porosity, surface cracks, and organic matter content.

Values of *K* also can be obtained for the soils that have been classified in a state by contacting any Soil Conservation Service office. These values have been determined for undisturbed soils, but they should be reasonable estimates of the *K* value for the replaced topsoil.

SLOPE LENGTH AND SLOPE STEEPNESS FACTORS

The slope length (*L*) and slope steepness (*S*) factors can be considered together as a single factor (*LS*). Table 26 or Figure 51 can be used to obtain this *LS* factor for uniform slopes. Irregular slopes require other methods for determining the *LS* value, but most of the reclaimed surface mines will have relatively uniform slopes and Table 26 or Figure 51 can be used to determine the *LS* value.

If the slope is not uniform or straight, it will be concave or convex. The slope can also be composed of a combination of concave and convex segments. These irregular slopes can be divided into segments of equal length that are reasonably uniform, and Tables 26 and 27 can be used to determine the *LS* value for each segment. These values can then be added to determine the *LS* value for the entire slope.

To evaluate an irregular slope, the slope is divided into segments that are of approximately equal length with each segment being approximately straight or uniform. As an example, let us assume a convex slope has a 3 percent slope at the top fourth of the slope length, 4 percent at the next fourth, 6 percent at the next fourth, and 8 percent at the lowest fourth. The length of the entire slope is 400 feet. A table for calculating the *LS* value of the entire slope would be as follows:

Sequence No. of Segment	Segment Length (ft)	Percent Slope	Value from Table 26[1]		Value from Table 27		Product
1	100	3	0.437	x	0.17	=	0.074
2	100	4	0.697	x	0.24	=	0.167
3	100	6	1.350	x	0.30	=	0.425
4	100	8	1.980	x	0.35	=	0.693
			LS of entire slope			=	1.359

[1]Note that entire length of slope is used to locate values in Table 26.

There may be different types of soil on a slope. If this happens the effect of this variation can be calculated along with the variation in slope. Assume that a slope is 300 feet long, concave, and can be divided

Table 27. Estimated Relative Soil Losses from Successive Equal-Length Segments with Uniform Slopes[1]

No. of segments	Sequence no. of segment	Fraction of soil loss		
		$m = 0.5$	$m = 0.4$	$m = 0.3$
2	1	0.35	0.38	0.41
	2	.65	.62	.59
3	1	.19	.22	.24
	2	.35	.35	.35
	3	.46	.43	.41
4	1	.12	.14	.17
	2	.23	.24	.24
	3	.30	.29	.28
	4	.35	.33	.31
5	1	.09	.11	.12
	2	.16	.17	.18
	3	.21	.21	.21
	4	.25	.24	.23
	5	.28	.27	.25
6	1	.07	.08	.10
	2	.12	.13	.14
	3	.16	.16	.17
	4	.19	.19	.18
	5	.22	.21	.20
	6	.24	.22	.21

[1]Derived by the formula:

$$\text{soil loss fraction} = \frac{\underline{i}^{\underline{m}+1} - (\underline{i}-1)^{\underline{m}+1}}{\underline{N}^{\underline{m}+1}}$$

where $\underline{i}$ = segment sequence number; $\underline{m}$ = slope length exponent (0.5 for slopes > 5 percent, 0.4 for 4 percent slopes, and 0.3 for 3 percent or less); and $\underline{N}$ = number of equal length segments into which the slope was divided.

Source: Wischmeir, W.H., and Smith, D.D. 1978. Predicting Rainfall Erosion Losses: A Guide to Conservation Planning. Agriculture Handbook No. 537, U.S. Dept. of Agriculture, Science and Education Administration. Washington, D.C.: U.S. Government Printing Office.

into three segments of approximately uniform shape. Segment 1 has a slope of 8 percent and a soil with a K value of 0.22. The second segment has a slope of 6 percent and the top half has a K value of 0.22, whereas the bottom half has a K value of 0.31. The bottom segment has a slope of 3 percent and a K value of 0.31. The complete slope is evaluated for a KLS value in the following manner.

Sequence No. of Segment	Segment Length (ft)	Percent Slope	Value from Table 26		Value from Table 27		K		Product
1a	50	8	1.72	x	0.07	x	0.22	=	0.026
1b	50	8	1.72	x	0.12	x	0.22	=	0.045
2a	50	6	1.17	x	0.16	x	0.22	=	0.041
2b	50	6	1.17	x	0.19	x	0.31	=	0.069
3	50	3	0.40	x	0.20	x	0.31	=	0.025
3b	50	3	0.40	x	0.21	x	0.31	=	0.026
					KLS of entire slope			=	0.232

VEGETATIVE FACTOR

When topsoil has been replaced on a regraded overburden, the topsoil has no protection from the erosive effects of rainfall. Under these conditions the vegetative factor (*C*) has a value of 1.0. However, the application of a mulch will immediately reduce this value to a more acceptable level. Figure 52 shows the degree of cover expected from various amounts of straw or hay mulch. Figure 53 shows the *C* values expected from various degrees of soil cover resulting from mulching. Figures 52 and 53 together can be used to determine the *C* value

Figure 52. Relation of percent cover to dry weight of uniformly distributed straw or hay mulch. *Source:* Wischmeir, W.H., and Smith, D.D. 1978. *Predicting Rainfall Erosion Losses.* U.S.D.A. Agr. Handbook 537.

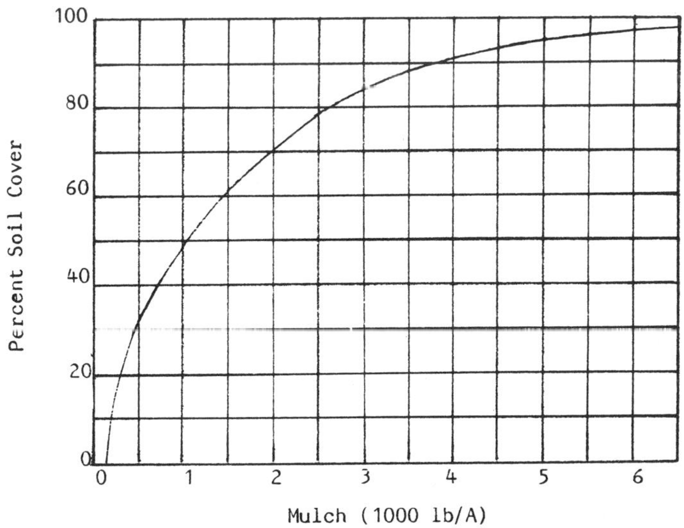

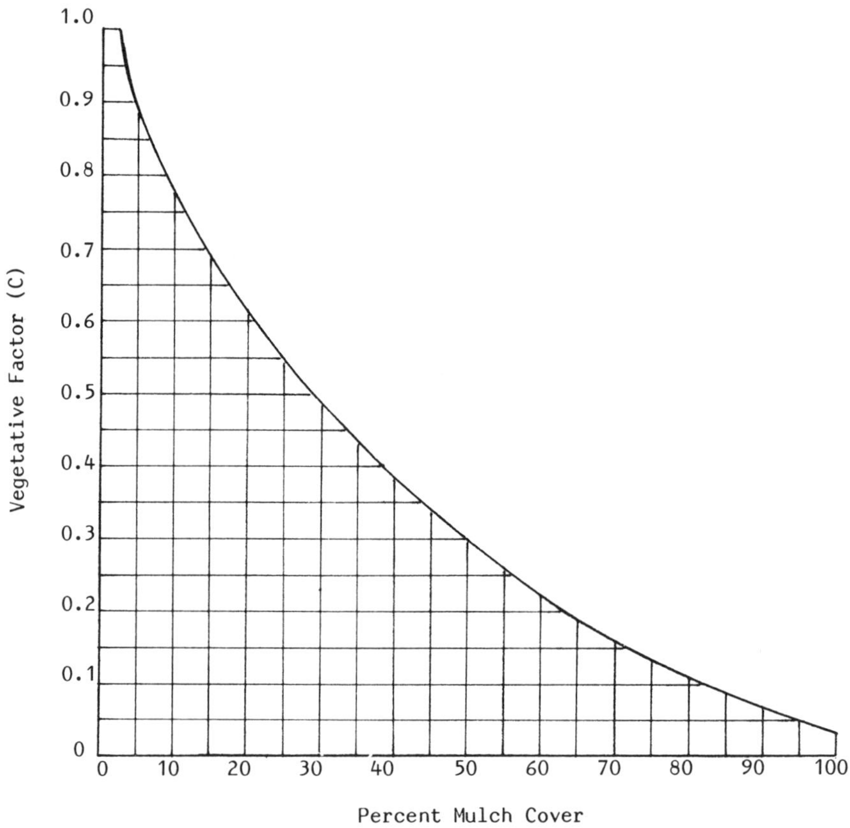

Figure 53. Effect of mulch on vegetative factor (C). *Source:* Wischmeir, W.H., and Smith, D.D. 1978. *Predicting Rainfall Erosion Losses.* U.S.D.A. Agr. Handbook 537.

expected as a result of the amount of straw or hay applied to the topsoil. These *C* values apply only for straw or hay mulches that have been secured by asphalt emulsions, discing, crimping, or some other method of anchoring the mulch.

There are mulching materials other than straw or hay and the *C* values for a few of these are shown in Table 28. These values are given for slopes of varying percents and lengths.

MECHANICAL EROSION CONTROL FACTOR

The mechanical erosion control factor (*P*) is used to evaluate the erosion control effectiveness of such measures as contour plowing, terracing, and diversion ditches. These practices become more important as slopes

Table 28. Vegetative Factor (*C*) Values and Slope Length Limits for Various Mulches[1]

Not seeded[2]				
	Slope[3]			Maximum length[4]
Mulch	**(ton/A)**	**(percent)**	***C* value**	**(ft)**
1. No mulching or seeding		All	1.0	—
2. Straw or hay, tied down by anchoring or tacking equipment used on slope[5]	1.0	5	0.20	200
		6–10	0.20	100
	1.5	5	0.12	300
		6–10	0.12	150
	2.0	5	0.06	400
		6–10	0.06	200
		11–15	0.07	150
		16–20	0.11	100
		21–25	0.14	75
		26–33	0.17	50
		34–50	0.20	35
3. Crushed stone (0.25 to 1.50 in.)	135	15	0.05	200
		16–20	0.05	150
		21–33	0.05	100
		34–50	0.05	75
	240	20	0.02	300
		21–33	0.02	200
		34–50	0.02	150
4. Wood chips	7	15	0.08	75
		16–20	0.08	50
	12	15	0.05	150
		16–20	0.05	100
		21–33	0.05	75
	25	15	0.02	200
		16–20	0.02	150
		21–33	0.02	100
		34–50	0.02	75

Seeded			
		***C* values[6]**	
Seeding	**Mulch (ton/A)**	**Pd 1**	**Pd 2**
Temporary (grain or fast-growing grass)	None	0.70	0.10
	Straw (1)	0.20	0.07
	Straw (1.5)	0.12	0.05

continued

Table 28. Vegetative Factor (*C*) Values and Slope Length Limits for Various Mulches[1] *(Continued)*

Seeded			
Seeding	**Mulch (ton/A)**	**C values[6]**	
		Pd 1	**Pd 2**
	Straw (2)	—[7]	0.05
	Stone (135)	0.05	0.05
	Stone (240)	0.02	0.02
	Wood chips (7)	0.08	0.05
	Wood chips (12)	0.05	0.02
	Wood chips (25)	0.02	0.02
Permanent seeding, second year		—	0.01
Sod		0.01	0.01

[1]Based on research data and field experience; prepared at a workshop of personnel from U.S.D.A. Agricultural Research Service, Soil Conservation Service, and various Maryland state and local agencies.

[2]If seeding is late in fall, these values would extend until the next spring.

[3]Slopes steeper than 50 percent not considered.

[4]Maximum length of slope for which mulch application is permitted. If exceeded, a greater application rate is required or the slope should be shortened with benches, diversions, etc.

[5]If the straw is not anchored to the soil, rilling may occur beneath the mulch. C values on moderate or steep slopes of soils having a K value greater than 0.3 should then be taken at double the values shown here.

[6]Pd 1 = period 1: through first 6 weeks of growing period. Pd 2 = period 2: after 6 weeks of growth.

[7]Use values for no seeding for appropriate slope steepness.

become steeper and longer. Placing a diversion across the middle of a slope that is 200 feet long effectively reduces the slope to 100 feet in length. This will greatly reduce the amount of soil erosion that could be expected. It is obvious that terraces and diversions must be taken into account when applying the erosion equation because they change the *LS* factor.

If no mechanical erosion control structures are used, the factor (*P*) will have a value of 1.0. If terraces or diversion ditches are constructed, they will produce the *P* values shown for different slopes in Table 29.

Table 29. Values for Mechanical Erosion Control Factor

Slope (percent)	P
1-2	0.12
3-8	0.10
9-12	0.12
13-16	0.14
17-20	0.16
21-25	0.18

USING THE UNIVERSAL SOIL LOSS EQUATION

The following assumptions regarding a hypothetical reclamation situation can be used to illustrate some uses of the universal soil loss equation. Assume that the mine is located in the northwestern quarter of Alabama; the replaced topsoil is 60 percent silt plus very fine sand, 20 percent sand, 0 percent organic matter; and that it has a fine granular structure and moderate permeability. The slope is 225 feet long, uniform, and has a gradient of 20 percent. There is no vegetation on the topsoil and no water control structures such as diversion ditches are constructed. Under these conditions the values for the soil loss equation will be:

$$E = RKLSCP$$

E = tons of soil eroded per acre per year

R = 325 (interpolated from Figure 49)

K = 0.42 (from Figure 50)

LS = 6.10 (from Figure 51)

C = 1.00 (when no mulch or other vegetation present)

P = 1.00 (when no control structure used)

E = 325 x 0.42 x 6.10 x 1.00 x 1.00 = 832.65 ton/A/yr

A 6-inch thick layer of soil covering one acre weighs approximately 1000 tons. Therefore, a 6-inch layer of topsoil would be completely eroded in 1 year and 73 days under these circumstances. This equation can also be solved by applying the nomograph in Appendix D if the C and P factors are both equal to 1.0.

Table 30. Percentage of Total Erosion That Normally Occurs Between January 1 and the Indicated Dates.[1] Computed for the Geographic Areas Shown in Figure 54.

Area no.	Jan. 1	Jan. 15	Feb. 1	Feb. 15	Mar. 1	Mar. 15	Apr. 1	Apr. 15	May 1	May 15
1	0	0	0	0	0	0	1	2	3	6
2	0	0	0	0	1	1	2	3	6	10
3	0	0	0	0	1	1	2	3	6	13
4	0	0	1	1	2	3	4	7	12	18
5	0	1	2	3	4	6	8	13	21	29
6	0	0	0	0	1	1	1	2	6	16
7	0	1	1	2	3	4	6	8	13	25
8	0	1	3	5	7	10	14	20	28	37
9	0	2	4	6	9	12	17	23	30	37
10	0	1	2	4	6	8	10	15	21	29
11	0	1	3	5	7	9	11	14	18	27
12	0	0	0	0	1	1	2	3	5	9
13	0	0	0	1	1	2	3	5	7	12
14	0	0	0	1	2	3	4	6	9	14
15	0	0	1	2	3	4	6	8	11	15
16	0	1	2	3	4	6	8	10	14	18
17	0	1	2	3	4	5	6	8	11	15
18	0	1	2	4	6	8	10	13	19	26
19	0	1	3	6	9	12	16	21	26	31
20	0	2	3	5	7	10	13	16	19	23
21	0	3	6	10	13	16	19	23	26	29
22	0	3	6	9	13	17	21	27	33	38
23	0	3	5	7	10	14	18	23	27	31
24	0	3	6	9	12	16	20	24	28	33
25	0	1	3	5	7	10	13	17	21	24
26	0	2	4	6	8	12	16	20	25	30
27	0	1	2	3	5	7	10	14	18	22
28	0	1	3	5	7	9	12	15	18	21
29	0	1	2	3	4	5	7	9	11	14
30	0	1	2	3	4	5	6	8	10	14
31	0	0	0	1	2	3	4	5	7	12
32	0	1	2	3	4	5	6	8	10	13
33	0	1	2	4	6	8	11	13	15	18

June		July		Aug.		Sept.		Oct.		Nov.		Dec.	
1	15	1	15	1	15	1	15	1	15	1	15	1	15
11	23	36	49	63	77	90	95	98	99	100	100	100	100
17	29	43	55	67	77	85	91	96	98	99	100	100	100
23	37	51	61	69	78	85	91	94	96	98	99	99	100
27	38	48	55	62	69	76	83	90	94	97	98	99	100
37	46	54	60	65	69	74	81	87	92	95	97	98	99
29	39	46	53	60	67	74	81	88	95	99	99	100	100
40	49	56	62	67	72	76	80	85	91	97	98	99	99
48	56	61	64	68	72	77	81	86	89	92	95	98	99
43	49	54	58	62	66	70	74	78	82	86	90	94	97
38	47	53	57	61	65	70	76	83	88	91	94	96	98
35	41	46	51	57	62	68	73	79	84	89	93	96	98
15	27	38	50	62	74	84	91	95	97	98	99	99	100
19	33	48	57	65	74	82	88	93	96	98	99	100	100
20	28	39	52	63	72	80	87	91	94	97	98	99	100
22	31	40	49	59	69	78	85	91	94	96	98	99	100
25	34	45	56	64	72	79	84	89	92	95	97	98	99
20	28	41	54	65	74	82	87	92	94	96	97	98	99
34	42	50	58	63	68	74	79	84	89	93	95	97	99
37	43	50	57	64	71	77	81	85	88	91	93	95	97
27	34	44	54	63	72	80	85	89	91	93	95	96	98
33	39	47	58	68	75	80	83	86	88	90	92	95	97
44	49	55	61	67	71	75	78	81	84	86	90	94	97
35	39	45	53	60	67	74	80	84	86	88	90	93	95
38	43	50	59	69	75	80	84	87	90	92	94	96	98
27	33	40	46	53	61	69	78	89	92	94	95	97	98
35	41	47	56	67	75	81	85	87	89	91	93	95	97
27	32	37	46	58	69	80	89	93	94	95	96	97	99
25	29	36	45	56	68	77	83	88	91	93	95	97	99
17	22	31	42	54	65	74	83	89	92	95	97	98	99
19	26	34	45	56	66	76	82	86	90	93	95	97	99
17	24	33	42	55	67	76	83	89	92	94	96	98	99
17	22	31	42	52	60	68	75	80	85	89	92	96	98
21	26	32	38	46	55	64	71	77	81	85	89	93	97

[1]For dates not listed in the table, interpolate between adjacent values.

Source: Wischmeir, W.H., and Smith, D.D. 1978. Predicting Rainfall Erosion Losses: A Guide to Conservation Planning. Agriculture Handbook No. 537, U.S. Dept. of Agriculture, Science and Education Administration. Washington, D.C.: U.S. Government Printing Office.

If the reclamationist decides better erosion control is needed, straw mulch can be applied to the topsoil at the rate of 1 ton per acre, with the straw anchored by crimping or applying asphalt. This action will provide a C value of 0.15 (Figures 52 and 53), and the soil loss equation now indicates the following soil loss:

$$E = 325 \times 0.42 \times 6.10 \times 0.15 \times 1.0 = 124.90 \text{ ton/A/yr}$$

Under these conditions it would take 8 years and 2 days to erode 6 inches of topsoil. Of course, this is a hypothetical situation because the straw would not remain in an unchanged condition for more than a few months.

The reclamationist may decide that the topsoil needs more protection and so will construct diversion ditches in order to reduce the slope lengths to 75 feet. This would require two ditches across the 225 feet of slope. This action will provide a P factor of 0.16 (Table 29), and the LS value for a 20 percent slope 75 feet in length is 3.53. The soil loss equation now indicates the following loss:

$$E = 325 \times 0.42 \times 3.53 \times 0.15 \times 0.16 = 11.56 \text{ ton/A/yr}$$

The addition of the diversion ditches increases the time required for loss of the 6 inches of topsoil to 86.51 years.

The above discussion is for the sake of examining the relative effects of the various factors controlling erosion. In a practical situation, the conditions under which erosion was calculated would change within the year. For instance, the topsoil would probably be seeded and vegetation would become established during the period of 1 year that was used in the preceding erosion calculations. If the minesoil is seeded with perennial grasses and legumes, and weather conditions are favorable, a ground cover of 80 or 90 percent may be obtained within the first year. A ground cover of 80 percent would provide a C value of 0.013, with the soil loss equation indicating the following loss:

$$E = 325 \times 0.42 \times 3.53 \times 0.013 \times 0.16 = 1.00 \text{ ton/A/yr}$$

The reclamationist would probably like to make estimates of the amount of erosion likely for each month within the year under different reclamation conditions. All of the erosion factors that have been discussed will probably remain constant from month to month except for the rainfall factor (R) and vegetation factor (C). Therefore, the erosion that can be expected for each month or other period can be calculated by taking these changes into account.

The percents of the total erosion that will have occurred on given dates during the year are shown in Table 30. These percents have been obtained for different sections of the country and these sections are shown in Figure 54. Figure 54 and Table 30 can be used to determine the percent of total erosion that will have occurred during any part of the year. As an example, assume that a mine is located in eastern Kentucky and a value for the percent of erosion that will occur by June 15 is needed. Figure 54 shows that eastern Kentucky is located in area

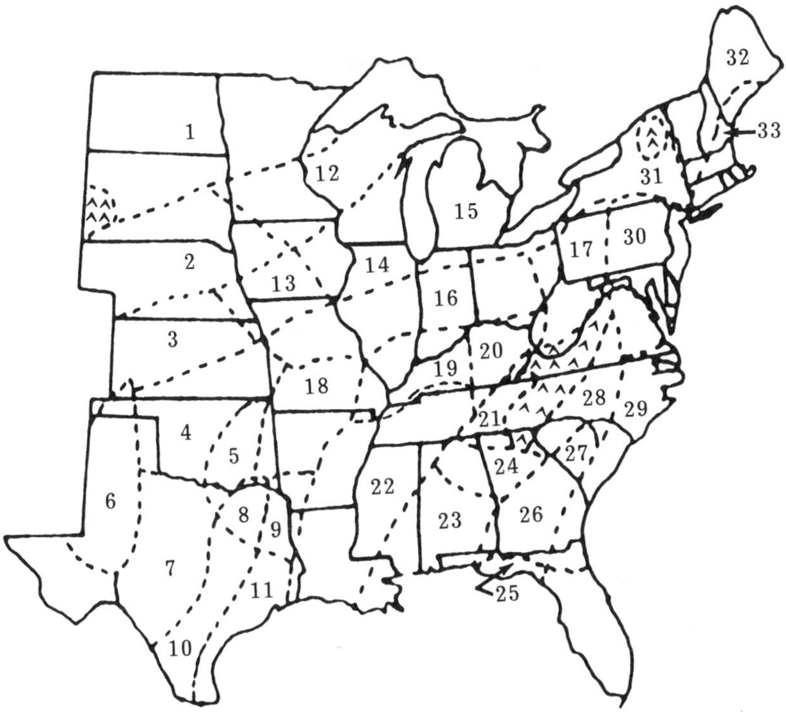

Figure 54. Map for area location for Table 30. *Source:* Wischmeir, W.H., and Smith, D.D. 1978. *Predicting Rainfall Erosion Losses: A Guide to Conservation Planning.* Agriculture Handbook No. 537, U.S. Dept. of Agriculture, Science and Education Administration. Washington, D.C.: U.S. Government Printing Office.

number 20. Table 30 gives a value of 34 percent for area 20 on June 10. This means that 34 percent of the yearly erosion will have occurred by that date.

Since the amount of rainfall and erosion varies from month to month during the year, it is imperative that it be taken into account when calculating yearly erosion. It is also important to take this variation into account when performing reclamation operations. For instance, it is better to replace topsoil during a less rainy period. If the topsoil had to be replaced during a rainy month, the application of mulch and construction of diversions could be considered in order to reduce erosion.

There is also the problem that revegetation begins with a bare soil that must be planted in order to obtain permanent erosion control. This bare soil period is the most erodible phase of the revegetation operation. If mulch is applied after seeding, the erosion potential is immediately and drastically reduced. If no mulch is applied, there is a period while the seeds are germinating and the seedlings becoming established when the soil can erode at a peak rate.

There is a method of taking these factors into account and obtaining a *C* value for the year that averages the effects of the varying crop stages and rainfall differences.

To use this method the information from Figure 54 and Table 30 are needed, plus an estimate of the soil loss expected during the stages of crop development from seeding to mature crop. Table 28 lists some *C* values and slope length limits for various mulches with and without seeding.

The *C* value for a particular reclamation situation can be estimated for a given set of circumstances. As an example, consider a situation where a mine is located in eastern Kentucky, the topsoil is replaced on March 1, a grass-legume mixture is sown on April 1, and no mulch is applied. Eastern Kentucky is in area 20 according to Figure 54.

The *C* value for the period from March 1 to April 1 will be 1.0, as the soil is bare during this period. From April 1 to May 15 the grass and legume seed are becoming established in period 1. According to Table 28, the *C* value for this period with no mulch is 0.70. The period from May 15 to October 1 is period 2 and the *C* value will be 0.10. The crop should be well established by October 1 and the *C* value should be 0.01 as long as the grass and legume mixture is maintained.

Date	Operation	*C*[1]		Parts of[2] Total Erosion		Partial *C*
3/1-4/1	Topsoil replacement	1.00	x	0.06	=	0.0600
4/1-5/15	Period 1	0.70	x	0.10	=	0.0700
5/15-10/1	Period 2	0.10	x	0.66	=	0.0660
10/1-3/1	Mature crop	0.01	x	0.18	=	0.0018
				Total *C*	=	0.1978

[1]From Table 28.
[2]From Table 30.

This value for *C* may now be used in the universal soil loss equation to estimate the yearly soil loss expected under these circumstances.

CHAPTER 15

MECHANICAL SOIL EROSION CONTROL

Diversion ditches and terraces are frequently used to control the erosive effects of rainfall on sloping soils. Diversion ditches have steep-sided channels that are difficult for farm machinery to traverse. These ditches can be as effective as terraces for erosion control, but they cause a loss of time and cultivated area if the land is to be cropped. Many terraces can be built in a manner that controls water movement but still allows farm equipment to move across them freely. Under these circumstances the area composed of terraces can be cropped as well as the area between the terraces.

Most surface mines will be reclaimed as pasture or forest land and will not require extensive machine operation for cultivation purposes; therefore the less expensive diversion ditch should be acceptable. Also, terraces often require deeper soil movement than diversion ditches and this may cause problems where the replaced topsoil is shallow.

TERRACES

Terraces are classified by applying three different terrace characteristics. These characteristics are: (1) the shape of the terrace cross section; (2) the way in which the runoff is handled; and (3) the alignment between terraces.

Terrace Characteristics

Shape	Channel	Alignment
Broadbase	Graded	Parallel
Steep backslope	Level	Nonparallel
Flat channel		
Bench		

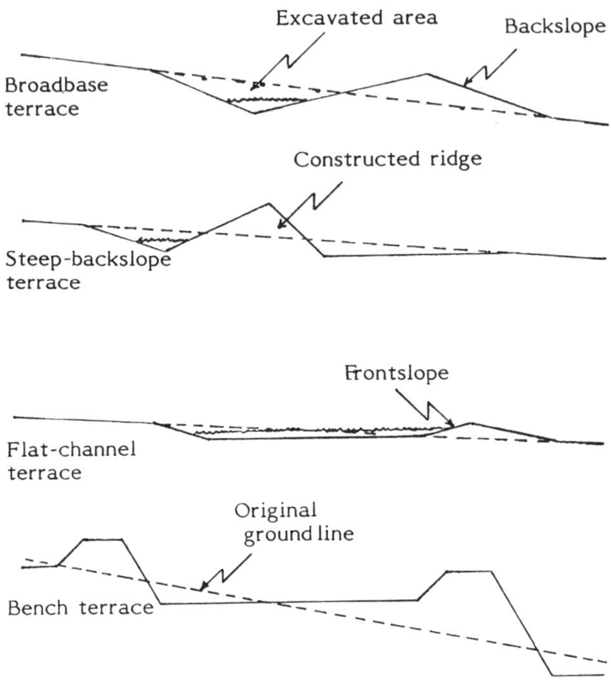

Figure 55. Terrace cross sections.

There are four general terrace cross sections. Figure 55 shows the original ground line, excavated area, constructed ridge, front slope, back slope, and approximate shape of the cross section for each of these terraces.

If the terrace is constructed with a gradient that allows runoff to move down the terrace channel, it is called a graded terrace. If the terrace is constructed without any gradient, it is called a level terrace. The graded terrace is used to move water from a slope without eroding the slope. If the terrace is constructed in an area where rainfall is not excessive, a level terrace with closed ends can be used. This type of terrace conserves water and allows it to percolate into the soil rather than lose it as runoff. Level terraces can also be constructed with open ends if rainfall is too great for closed ends but not great enough for a graded terrace.

The alignment between terraces can be parallel or nonparallel. Parallel terraces are used where farming operations require extensive machine operations that will be facilitated by long parallel rows. Nonparallel terraces are used for crops such as forage or small grains that do not require frequent machine operations.

To completely identify a terrace system, all three characteristics must be used. For instance, a system could be described as a broadbase, graded, and parallel system or a flat-channel, level, and nonparallel system.

The broadbase terrace is used when the entire area is to be cultivated and the slope is less than 12 percent. This terrace can be level or graded and parallel or nonparallel. Soil to build the terrace is taken from the uphill side and this tends to increase the slope between terraces. If soil were taken from the downhill side, the slope between terraces would be reduced.

The steep-backslope terrace is indicated for areas that are too steep for broadbase terraces. The backslope of these terraces is vegetated with grass, shrubs, or trees and left undisturbed to prevent erosion. These terraces can be graded or level and parallel or nonparallel. Soil may be taken from the downhill side or from the uphill and downhill sides.

The flat-channel terrace may have a channel as wide as 75 feet. These terraces are used where the annual rainfall is less than 25 inches. As the channel is wide, it can be used to store water and allow it to percolate into the soil. To facilitate this conservation, the terrace is usually level and closed at each end. These terraces may be parallel or nonparallel, and soil is usually taken from the uphill side to construct the terrace.

The uphill and downhill slopes of the bench terrace are too steep for cultivation. However, a wide level bench is formed on the uphill side that can be cultivated. Soil is removed on the downhill side and used to form the wide level bench with a steep ridge on the downhill edge of the bench. These terraces can be parallel or nonparallel but are usually constructed with a grade parallel to the backslope.

DIVERSION DITCHES

Diversion ditches are usually considered to be single structures used to intercept a surface water flow and divert it away from an area to be cultivated. A terrace system is then constructed below the diversion ditch to prevent erosion and facilitate soil water infiltration on the cultivated area. However, diversion ditches as used on most surface mines are similar in construction to broadbase terrace systems. The only differences are that the channel slopes and backslopes are usually steeper on the diversion ditches. These ditches are used in the same way on surface mines as terraces are used on cultivated fields. It is difficult to cross the slopes formed by diversion ditches, but most surface mines are planted with forage crops and there is little need to cross the steep slopes.

Soil scientists are aware of principles and have research results that can be used to evaluate many of the factors that can be manipulated to control soil erosion. As more of these factors are taken into account, the chances of controlling erosion become greater. However, evaluating all of these erosion factors requires time and expense. There are many times when the reclamationist needs some quick guides to design a diversion ditch system that will most likely ameliorate the immediate erosion problem. The following paragraphs describe a simplified method of designing diversion ditch systems. To design the system, several

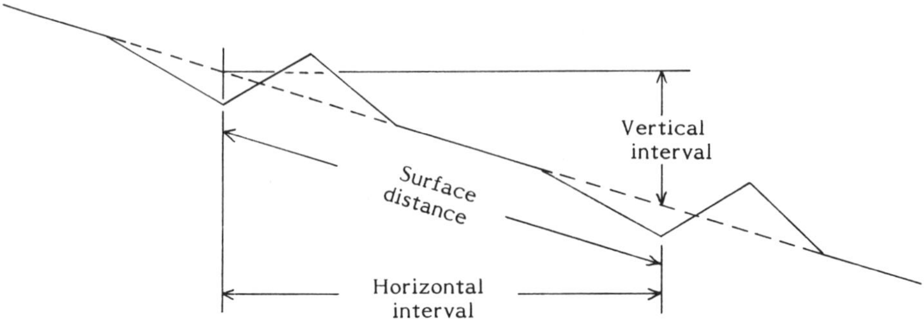

Figure 56. Methods of measuring distance between diversion ditches or terraces.

pieces of information are needed; for instance, slope of the land, rainfall intensity, soil characteristics, and type and amount of vegetation present or expected.

SPACING

Spacing between ditches is described as vertical or horizontal (Figure 56). Either of these methods can be easily applied with a hand level or more sophisticated level. The horizontal interval can be measured as the distance along the ground surface, without considering slope until the slope reaches approximately 20 percent. Horizontal distances must be measured when the slope exceeds 20 percent.

The factors that control the amount of erosion on a slope are the steepness of slope, length of slope, rainfall intensity, soil characteristics, and soil cover. The following equation takes most of these factors into account and was developed as a guide for estimating the spacing needed between terraces and diversion ditches.

$$VI = XS + Y$$

VI = vertical interval or difference in elevation between ditches (feet)

X = value obtained from Figure 57

S = average slope of the land draining into the ditch in feet per 100 feet (percent slope)

Y = variable value determined by soil erodibility and soil vegetative cover

The values for Y are described by the Soil Conservation Service as follows:

Y = 1.0 for soils with below-average water intake rates and vegetation that provides little cover during intense rainfall periods. Much of the rainfall will run off this soil and erosion will be great.

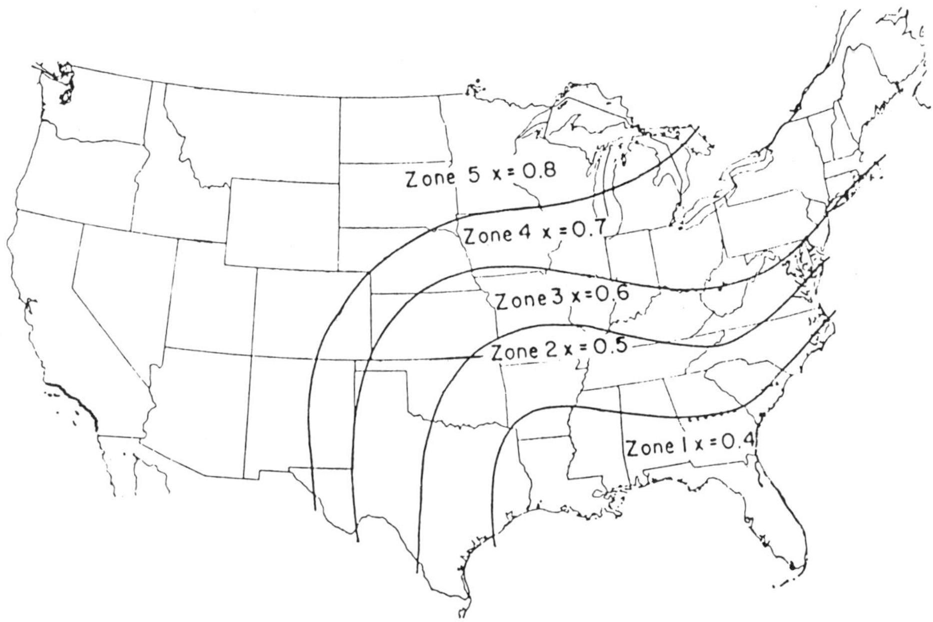

Figure 57. Values of *X* for the equation $VI = XS + Y$. *Source:* U.S. Department of Agriculture, Soil Conservation Service. 1979. S.C.S. Engineering Field Manual for Conservation Practices. Washington, D.C.: U.S. Government Printing Office.

$Y = 4.0$ for soils with average or above-average water intake rates and vegetation that provides good cover during periods of intense rainfall. A small percent of the rainfall will run off this soil and erosion will be slight.

$Y = 2.5$ where one of the above factors is favorable and the other unfavorable.

When the topsoil is high in silt and very fine sand and no mulch is applied, the *Y* value should probably be 1.0. However, if the topsoil is high in coarse sand and clay *or* if at least 1 ton per acre of mulch is applied, the *Y* value could be raised to 2.5. If the topsoil is high in coarse sand and clay *and* 1 ton or more of mulch is applied per acre, the *Y* value can be raised to 4.0.

Example: Assume that a mine is located in southern Illinois and the slope of the regraded and topsoiled mine is 12 percent. The topsoil has a high silt content and 1 ton per acre of mulch is applied.

$VI = XS + Y$

$X = 0.5$ (from Figure 57)

$S = 12$

$Y = 2.5$ (because the silt will probably be highly erodible but the mulch will prevent excessive erosion)

$VI = (0.5 \times 12) + 2.5$

$= 6.0 + 2.5 = 8.5$ feet

It must be remembered that values obtained with this equation are estimates. It seems reasonable to vary the spacing values obtained by as much as 10 percent and still expect good erosion control.

When the vertical distance is used to space terraces or diversions, the changes in slope will cause the terraces or diversions to be non-parallel. If parallel terraces or diversions are desired the horizontal distance must be calculated by using the previously calculated vertical distance. The equation for calculating horizontal interval or distance is as follows:

$$HI = \frac{VI}{S} \times 100$$

HI = horizontal interval in feet

VI = vertical interval in feet

S = percent slope

Example: Use the vertical interval calculated in the previous example and convert vertical interval (VI) to horizontal interval (HI).

$$HI = \frac{VI}{S} \times 100$$

$$HI = \frac{8.5}{12} \times 100$$

$$HI = 70.0 \text{ feet}$$

The horizontal distance can be considered the same as surface distance for slopes up to approximately 20 percent. This will save time when measuring the distance between terraces or diversions. The calculated horizontal distance can now be used as a constant value in the following equation to calculate vertical distances that will result in parallel terraces or diversions.

$$VI = \frac{HI \times S}{100}$$

The uppermost diversion should be established first because the slope ridge or drainage divide is the logical place to begin measuring. Measure the vertical, horizontal, or surface distance from the ridge to the first diversion and then measure from each downhill diversion to the next downhill diversion.

Other equations have been developed for diversion ditch or terrace spacing and a more general equation has been developed by the Soil Conservation Service.

$$HI = \frac{200}{S} + 50$$

HI = horizontal interval or distance between diversions in feet

S = slope in percent

Example: Assume that the mine is located in northern West Virginia and the slope is 15 percent.

$$HI = \frac{200}{S} = 50$$

$$HI = \frac{200}{15} + 50$$

$$= 13 + 50$$

$$= 63 \text{ feet}$$

Rainfall is more intense and there is a greater volume of rain per year in the southeastern U.S. Therefore, a constant of 25 should be used instead of 50 in order to reduce the distance between diversions and the effects of the more intense rainfall.

GRADIENT

Practically all diversions in the eastern U.S. should have a grade from the upper part of the diversion to the bottom where the runoff empties into an outlet channel. If the grade is too gentle, the diversion will not move the water fast enough and silting in the channel or an overflow of the diversion outer wall may occur. If the grade is too steep, the rushing water will scour the bottom and sides of the diversion channel.

The minimum grade for diversions in surface mine reclamation will probably be about 0.25 percent (0.25 foot of grade per 100 feet of ditch). It is difficult to lay out diversions with less grade when using hand levels. More sophisticated leveling devices and careful channel construction will allow lesser gradients to be used.

The maximum allowable grade depends on the erodibility of the soil and the depth of the water flow in the diversion ditch. The purpose of having a maximum allowable grade is to prevent an excessive water velocity that would erode the diversion ditch channel. Diversion ditches used on a mine will be bare of vegetation at the beginning and will need

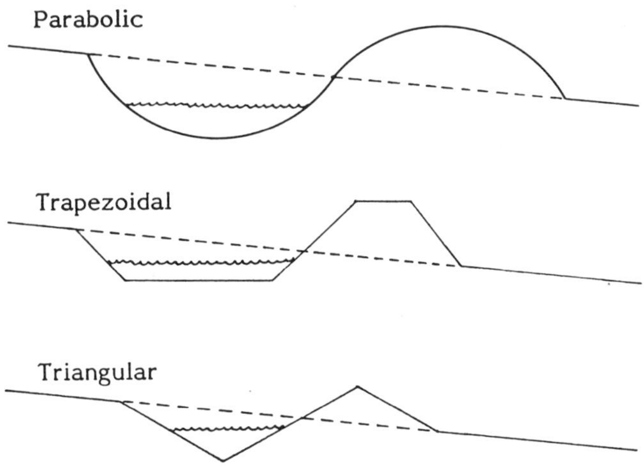

Figure 58. Parabolic, trapezoidal, and triangular cross sections for diversion ditch channels.

a small slope in order to obtain minimum erosion. As vegetation becomes established the slope could be greater. Under these circumstances the reclamationist may want to increase the slope above the calculated maximum allowable and accept the increased channel erosion until vegetation becomes established. This increased slope may be necessary when time and equipment prevent the construction of extremely gentle channel slopes. The maximum velocities that are usually accepted as safe are 2.5 feet per second for soils that are highly resistant to erosion and 1.5 feet per second for those highly susceptible to erosion. Therefore, many soils should be safe from erosion at a water velocity of 2.0 feet per second. Experience will soon teach the reclamationist the proper gradients for specific soils and conditions.

DIVERSION DITCH DESIGN

Diversion ditches on minesoils are usually designed to carry the peak runoff expected in a 10-year period for a 24-hour duration rainfall (see Chapter 13, Figure 41). These ditches can have any one of three cross-section shapes. The three shapes are parabolic, trapezoidal, and triangular (Figure 58). Any one of these shapes can be used to carry the peak runoff at a velocity that is slow enough to prevent serious erosion and yet fast enough to prevent the deposition of sediment in the channel bottom. The channel shape constructed will depend on such factors as steepness of land slope, desired velocity, available equipment, and minesoil characteristics. Diversions can be designed in such a way that they will have a cross section and gradient that will move all the water without scouring the channel bottom. However, this is often not necessary on overburdens because overburdens are often stony and resistant to erosion.

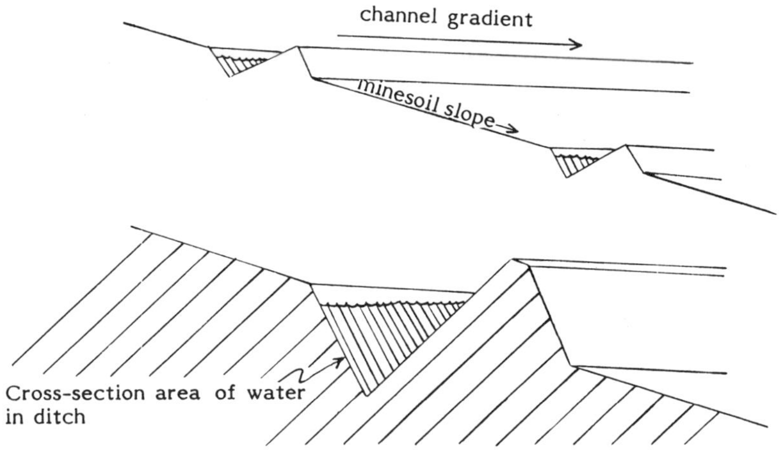

Figure 59. Channel gradient of diversion ditch, slope of minesoil, and cross-section area of water in diversion ditch channel.

Cross section has two characteristics that must be taken into account. The area of the cross section obviously has an effect on the amount of water moved. However, the shape of the cross section also has an effect on the speed or velocity with which water moves through the channel. The channel shape that contains the most water with the least amount of water contacting the channel sides is the channel that will move water at the greatest velocity. This is true because the friction produced by water flowing against the channel sides reduces the velocity of the water flow. Also, as the channel sides become rougher the velocity is further reduced. Channel roughness can be caused by many things such as rocks, grass, and unevenness of the channel surface. Channel gradient is the other factor to be determined and it obviously has an effect on velocity of water flow.

The quantity of water and the velocity of flow are controlled by all the above factors. Therefore, the reclamationist must be able to manipulate some or all of these factors in order to construct diversions that will control runoff in such a way that reclamation objectives can be obtained.

Two basic equations are needed in order to design diversions. The first equation is:

$$Q = V \times A$$

Q = quantity of water flowing in the channel in cubic feet per second (cfs)

V = average water velocity in feet per second (ft/s)

A = cross-sectional area of water flow in square feet (Figure 59)

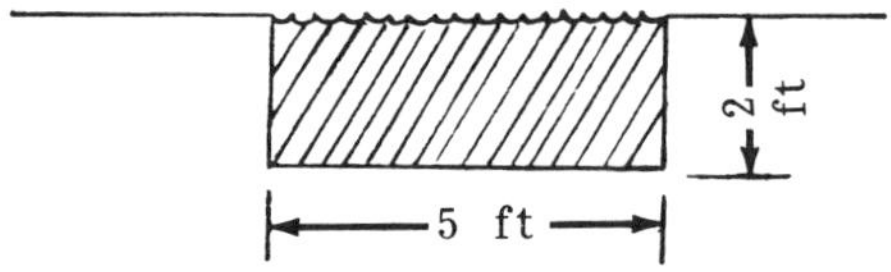

Area of this rectangular channel: 2 ft x 5 ft = 10 ft^2

Wetted perimeter when the channel is filled: 2 ft + 5 ft + 2 ft = 9 ft

Therefore, hydraulic radius: 10 ft^2 ÷ 9 ft = 1.11 ft

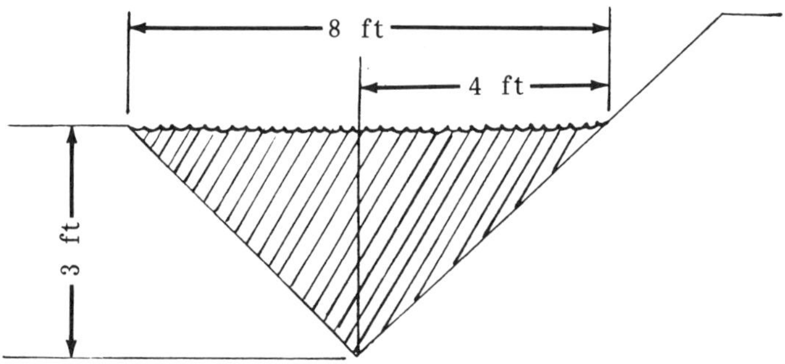

Area of this equilateral triangle: $\frac{3 \text{ ft} \times 8 \text{ ft}}{2} = 12 \text{ ft}^2$

Wetted perimeter when the channel is filled: $2\sqrt{3^2 + 4^2} = 10$ ft

Therefore, hydraulic radius: 12 ft^2 ÷ 10 ft = 1.2 ft

Figure 60. A graphic explanation of hydraulic radius.

The second equation is the Manning formula:

$$V = \frac{1.49}{n} R^{2/3} S^{1/2}$$

V = average velocity in feet per second

R = hydraulic radius in feet

S = gradient in *foot of drop per foot* of ditch length (Figure 59)

n = channel roughness coefficient (Table 31)

Hydraulic radius is the cross-sectional area of the channel divided by the wetted perimeter of the channel. Figure 60 offers a graphic explanation of hydraulic radius, and Figure 61 shows formulas for obtaining values needed in the Manning equation for triangular, trapezoidal, and parabolic cross sections.

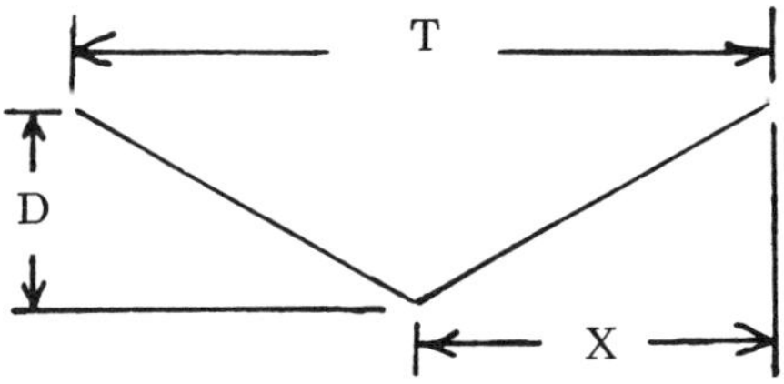

Slope (S) = $\frac{X}{D}$

Area	=	$\frac{DT}{2}$
Wetted perimeter	=	$2\sqrt{D^2+(\frac{T}{2})^2}$
Hydraulic radius	=	$\frac{\frac{DT}{2}}{2\sqrt{D^2+(\frac{T}{2})^2}}$
Top width	=	2DS

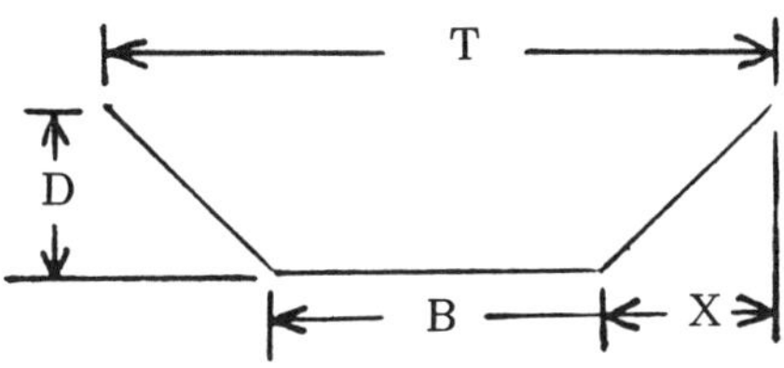

Slope (S) = $\frac{X}{D}$

Area	=	$BD + SD^2$
Wetted perimeter	=	$B + 2D\sqrt{1+S^2}$
Hydraulic radius	=	$\frac{BD + SD^2}{B + 2D\sqrt{1+S^2}}$
Top width	=	$T = B + 2DS$

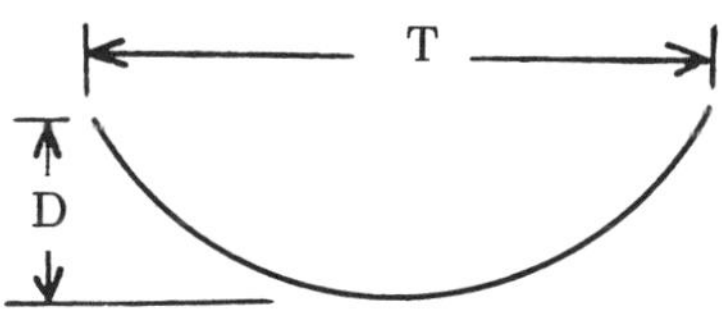

Area	=	$\frac{2}{3}TD$
Wetted perimeter	=	$T + \frac{8D^2}{3T}$
Hydraulic radius	=	$\frac{2T^2D}{3T^2+8D^2}$
Top width	=	$\frac{3(\text{area})}{2D}$

Figure 61. Formulas for obtaining values needed to solve the Manning equation for triangular, trapezoidal, and parabolic cross sections.

Table 31. Values of the Coefficient of Roughness (*n*) to be Used in Manning's Equation[1]

Type of channel	*n*
Lined	
Concrete, float finish	0.015
Concrete, trowel finish	0.013
Excavated soil	
Fairly straight, uniform cross section, free of vegetation	0.022
Fairly straight, uniform cross section, grass on sides of channel only	0.030
Not straight, cross section not uniform, stone or gravel on bottom	0.030
Natural	
Fairly straight, no vegetation	0.030
Not straight, no vegetation	0.035
Not straight, with vegetation	0.100

[1]A method for estimating the coefficient of roughness (n) is described in Appendix C.

The roughness coefficient has been determined for a number of situations (Table 31). It seems likely that the value of *n* for diversions on unvegetated minesoils will be close to 0.022. However, a thorough examination of the diversion channel could show that some other *n* value would be more valid. See Appendix C for a method of estimating *n* values.

TRIANGULAR CHANNEL DESIGN

The equation for peak runoff (*Q*) from a watershed or from the area between diversion ditches is used as a starting point in determining the cross-sectional area, shape, and gradient needed for a diversion ditch. The value of Q in the equation is obtained by applying Beasley's equation, the Soil Conservation Service method, or the rational formula as described in the runoff section.

Example: The Beasley equation is used in this example. The mine is located in the southeastern quarter of Ohio. The minesoil has a sandy loam texture. The average slope is 10 percent. The distance between ditches is 75 feet and they are 1000 feet in length. This yields an area of 1.7 acres between ditches. There is no vegetative cover, no contour farming, and no surface storage. The diversion ditch system is to be designed for the expected rainfall in a 10-year period for a 24-hour duration storm. Under these conditions the estimated peak or maximum runoff would be:

Table 32. Maximum Flow Velocities and Possible Gradients for Diversion Ditches with Different Soil Textures and Materials

Channel lining	Maximum velocity (ft/s)	S^1 (percent)	S^1 (ft/ft)
Sand and sandy loam	2.5	0.5	0.005
Silt loam	3.0	1.0	0.01
Sandy clay loam	3.5	1.5	0.015
Clay loam	4.0	2.0	0.02
Clay, fine gravel, and loam and gravel mixture	5.0	2.5	0.025
Silt and stone mixture	5.5	3.0	0.03
Shale, hardpan, and coarse gravel	6.0	3.5	0.035
Solid stone and cemented conglomerates	8.0	4.0	0.040

[1]Channel gradient.

$$Q = Q_T \times L \times I \times T \times S \times V \times C \times P \times F$$

$$Q = 6.5 \times 0.8 \times 0.9 \times 1.07 \times 0.8 \times 2.0 \times 1.0 \times 1.0 \times 1.0$$

$$Q = 8.0 \text{ cubic feet per second (ft}^3\text{/s) as runoff}$$

The maximum runoff velocity that could be tolerated without channel erosion on this soil would be 2.5 feet per second (Table 32). Now the equation can be solved for the cross-section area of the ditch.

$$Q = V \times A$$

$$A = \frac{Q}{V} = \frac{8 \text{ ft}^3\text{/s}}{2.5 \text{ ft/s}} = 3.2 \text{ ft}^2$$

The rest of this discussion is based on a channel with a cross section of equilateral triangular shape. To obtain the desired cross-section area for such a channel, the following equation is used:

$$A = \frac{D \times W}{2}$$

A = area of channel cross section in square feet

D = depth of channel in feet

W = maximum width of channel in feet

Choose a convenient depth or width and solve the equation for the other value. For this example, a depth of 1.5 feet has been chosen.

Table 33. Hydraulic Radii (R) for Triangular Ditches of Varying Depth and Width[1]

	Width									
Depth	**3**	**4**	**5**	**6**	**7**	**8**	**9**	**10**	**15**	**20**
0.5	0.237	0.243	0.245	0.247	0.247	0.248	0.248	0.249	0.249	0.250
0.75	0.335	0.351	0.359	0.364	0.367	0.369	0.370	0.371	0.373	0.374
1.0	0.416	0.447	0.464	0.474	0.481	0.485	0.488	0.490	0.496	0.498
1.5	0.530	0.600	0.643	0.671	0.689	0.702	0.712	0.718	0.735	0.742
2.0	0.600	0.707	0.781	0.832	0.868	0.894	0.914	0.928	0.966	0.981
2.5	0.643	0.781	0.894	0.960	1.017	1.060	1.093	1.118	1.186	1.213
3.0	0.671	0.832	0.960	1.061	1.139	1.200	1.248	1.286	1.393	1.437
3.5	0.689	0.868	1.017	1.139	1.237	1.317	1.381	1.434	1.586	1.652
4.0	0.702	0.894	1.060	1.200	1.317	1.414	1.495	1.562	1.765	1.857
4.5	0.712	0.914	1.093	1.248	1.381	1.495	1.591	1.672	1.929	2.052
5.0	0.718	0.928	1.118	1.286	1.434	1.562	1.672	1.768	2.080	2.236

[1]All data given in feet.

$$A = \frac{D \times W}{2}$$

$$3.2 \text{ ft}^2 = \frac{1.5 \text{ ft} \times W}{2}$$

$$W = \frac{2 \times 3.2}{1.5}$$

$$W = 4.27 \text{ ft}$$

The next step is to solve Manning's equation to find if this depth and width will provide the desired velocity of 2.5 feet per second. To solve this equation, the values of R (hydraulic radius), S (channel gradient), and n (roughness coefficient) must be obtained. In many cases the value of S can be selected for the convenience of the reclamationist. A value of 0.5 percent or 0.005 foot per foot will be selected in this instance. The value for n is selected from Table 31 and is likely to be 0.022 for newly constructed diversion ditches on a minesoil. The value for R can be calculated or obtained from Table 33.

Hydraulic radius for a channel with an equilateral triangular cross section can be calculated by applying the following equation:

$$R = \frac{A}{wp} = \frac{3.2}{2\sqrt{(1.5^2 + (2.13)^2}} = \frac{3.2}{5.21} = 0.614 \text{ ft}$$

R = hydraulic radius in feet

A = cross-section area in square feet

wp = wetted perimeter in feet

The value for $R^{2/3}$ can be calculated or obtained by interpolation from Table 34. The value for $S^{1/2}$ can also be calculated or obtained from Table 35. Manning's equation can now be solved for the velocity that would be developed at peak runoff.

$$V = \frac{1.49}{n} R^{2/3} S^{1/2}$$

$$V = \frac{1.49}{0.022} \times 0.722 \times 0.0707$$

V = 3.46 ft/s at 0.5 percent or 0.005 ft/ft gradient

Table 34. Values of $R^{2/3}$[1]

R[1]	.00	.01	.02	.03	.04	.05	.06	.07	.08	.09
.0	.000	.046	.074	.097	.117	.136	.153	.170	.186	.201
.1	.215	.229	.243	.256	.269	.282	.295	.307	.319	.331
.2	.342	.353	.364	.375	.386	.397	.407	.418	.428	.438
.3	.448	.458	.468	.477	.487	.497	.506	.515	.525	.534
.4	.543	.552	.561	.570	.578	.587	.596	.604	.613	.622
.5	.630	.638	.647	.655	.663	.671	.679	.687	.695	.703
.6	.711	.719	.727	.735	.743	.750	.758	.765	.773	.781
.7	.788	.796	.803	.811	.818	.825	.832	.840	.847	.855
.8	.862	.869	.876	.883	.890	.897	.904	.911	.918	.925
.9	.932	.939	.946	.953	.960	.966	.973	.980	.987	.993
1.0	1.000	1.007	1.013	1.020	1.027	1.033	1.040	1.046	1.053	1.059
1.1	1.065	1.072	1.078	1.085	1.091	1.097	1.104	1.110	1.117	1.123
1.2	1.129	1.136	1.142	1.148	1.154	1.160	1.167	1.173	1.179	1.185
1.3	1.191	1.197	1.203	1.209	1.215	1.221	1.227	1.233	1.239	1.245
1.4	1.251	1.257	1.263	1.269	1.275	1.281	1.287	1.293	1.299	1.305
1.5	1.310	1.316	1.322	1.328	1.334	1.339	1.345	1.351	1.357	1.362
1.6	1.368	1.374	1.379	1.385	1.391	1.396	1.402	1.408	1.413	1.419
1.7	1.424	1.430	1.436	1.441	1.447	1.452	1.458	1.463	1.469	1.474
1.8	1.480	1.485	1.491	1.496	1.502	1.507	1.513	1.518	1.523	1.529
1.9	1.534	1.539	1.545	1.550	1.556	1.561	1.566	1.571	1.577	1.582
2.0	1.587	1.593	1.598	1.603	1.608	1.613	1.619	1.624	1.629	1.634
2.1	1.639	1.645	1.650	1.655	1.660	1.665	1.671	1.676	1.681	1.686
2.2	1.691	1.697	1.702	1.707	1.712	1.717	1.722	1.727	1.732	1.737
2.3	1.742	1.747	1.752	1.757	1.762	1.767	1.772	1.777	1.782	1.787
2.4	1.792	1.797	1.802	1.807	1.812	1.817	1.822	1.827	1.832	1.837

continued

Table 34. Values of $R^{2/3}$[1] *(Continued)*

R[1]	.00	.01	.02	.03	.04	.05	.06	.07	.08	.09
2.5	1.842	1.847	1.852	1.857	1.862	1.867	1.871	1.876	1.881	1.886
2.6	1.891	1.896	1.900	1.905	1.910	1.915	1.920	1.925	1.929	1.934
2.7	1.939	1.944	1.949	1.953	1.958	1.963	1.968	1.972	1.977	1.982
2.8	1.987	1.992	1.996	2.001	2.006	2.010	2.015	2.020	2.024	2.029
2.9	2.034	2.038	2.043	2.048	2.052	2.057	2.062	2.066	2.071	2.075
3.0	2.080	2.085	2.089	2.094	2.099	2.103	2.108	2.112	2.117	2.122
3.1	2.126	2.131	2.135	2.140	2.144	2.149	2.153	2.158	2.163	2.167
3.2	2.172	2.176	2.180	2.185	2.190	2.194	2.199	2.203	2.208	2.212
3.3	2.217	2.221	2.226	2.230	2.234	2.239	2.243	2.248	2.252	2.257
3.4	2.261	2.265	2.270	2.274	2.279	2.283	2.288	2,292	2,296	2.301
3.5	2.305	2.310	2.314	2.318	2.323	2.327	2.331	2.336	2.340	2.345
3.6	2.349	2.353	2.358	2.362	2.366	2.371	3.375	2.379	2.384	2.388
3.7	2.392	2.397	2.401	2.405	2.409	2.414	2.418	2.422	2.427	2.431
3.8	2.435	2.439	2.444	2.448	2.452	2.457	2.461	2.465	2.469	2.474
3.9	2.478	2.482	2.486	2.490	2.495	2.499	2.503	2.507	2.511	2.516
4.0	2.520	2.524	2.528	2.532	2.537	2.541	2.545	2.549	2.553	2.558
4.1	2.562	2.566	2.570	2.574	2.579	2.583	2.587	2.591	2.595	2.599
4.2	2.603	2.607	2.611	2.616	2.620	2.624	2.628	2.632	2.636	2.640
4.3	2.644	2.648	2.653	2.657	2.661	2.665	2.669	2.673	2.677	2.681
4.4	2.685	2.689	2.693	2.698	2.702	2.706	2.710	2.714	2.718	2.722
4.5	2.726	2.730	2.734	2.738	2.742	2.746	2.750	2.754	2.758	2.762
4.6	2.766	2.770	2.774	2.778	2.782	2.786	2.790	2.794	2.798	2.802
4.7	2.806	2.810	2.814	2.818	2.822	2.826	2.830	2.834	2.838	2.842
4.8	2.846	2.850	2.854	2.858	2.862	2.865	2.869	2.873	2.877	2.881
4.9	2.885	2.889	2.893	2.897	2.901	2,904	2.908	2.912	2,916	2.920

[1]Hydraulic radius.

Another method of solving Manning's equation is by using the nomographs shown in Appendix E. The Soil Conservation Service has developed four of these nomographs to suit a wide range of values for *V, n, R,* and *S.* Values obtained from these nomographs are satisfactory approximations. The previous *n, R,* and *S* values yield a velocity of 3.45 ft/s when applied to Figure E-1 in Appendix E. To use this nomograph, a line is drawn from 0.614 on the *R* scale to 0.005 on the *S* scale. Next, a line is drawn from 0.022 on the *n* scale through the point on the dummy scale where the *R*-to-*S* line intersects the dummy scale. The line is continued to the *V* scale and the point on the *V* scale intersected by the line is the resulting velocity (3.45 ft/s) for the *n, R,* and *S* values. This velocity is greater than the desired velocity of 2.5 ft/s. Therefore, another attempt must be made to select the proper depth, width, and gradient. The

Table 35. Values of $S^{1/2}$[1]

S[1]	.–0	.–1	.–2	.–3	.–4	.–5	.–6	.–7	.–8	.–9
.001	.03162	.03317	.03464	.03606	.03742	.03873	.04000	.04123	.04243	.04359
.002	.04472	.04583	.04690	.04796	.04899	.05000	.05009	.05196	.05292	.05385
.003	.05477	.05568	.05657	.05745	.05831	.05916	.06000	.06083	.06164	.06245
.004	.06325	.06403	.06481	.06557	.06633	.06708	.06782	.06856	.06928	.07000
.005	.07071	.07141	.07211	.07280	.07348	.07416	.07483	.07550	.07616	.07681
.006	.07746	.07810	.07874	.07937	.08000	.08062	.08124	.08185	.08246	.08307
.007	.08367	.08426	.08485	.08544	.08602	.08660	.08718	.08775	.08832	.08888
.008	.08944	.09000	.09055	.09110	.09165	.09220	.09274	.09327	.09381	.09434
.009	.09487	.09539	.09592	.09644	.09695	.09747	.09798	.09849	.09899	.09950
.010	.10000	.10050	.10100	.10149	.10198	.10247	.10296	.10344	.10392	.10440
.01	.1000	.1049	.1095	.1140	.1183	.1225	.1265	.1304	.1342	.1378
.02	.1414	.1449	.1483	.1517	.1549	.1581	.1612	.1643	.1673	.1703
.03	.1732	.1761	.1789	.1817	.1844	.1871	.1897	.1924	.1949	.1975
.04	.2000	.2025	.2049	.2074	.2098	.2121	.2145	.2168	.2191	.2214
.05	.2236	.2258	.2280	.2302	.2324	.2345	.2366	.2387	.2408	.2429
.06	.2449	.2470	.2490	.2510	.2530	.2550	.2569	.2588	.2608	.2627
.07	.2646	.2665	.2683	.2702	.2720	.2739	.2757	.2775	.2793	.2811
.08	.2828	.2846	.2864	.2881	.2898	.2915	.2933	.2950	.2966	.2983
.09	.3000	.3017	.3033	.3050	.3066	.3082	.3098	.3114	.3130	.3146
.10	.3162	.3178	.3194	.3209	.3225	.3240	.3256	.3271	.3286	.3302

[1] Channel gradient.

Table 36. Cross-Section Areas of Ditches for Varying Runoffs and Velocities[1]

Runoff (Q)	Velocity (V) (ft/s)								
	1.0	**1.5**	**2.0**	**2.5**	**3.0**	**3.5**	**4.0**	**4.5**	**5.0**
5	5.0	3.3	2.5	2.0	1.7	1.4	1.2	1.1	1.0
10	10.0	6.7	5.0	4.0	3.3	2.9	2.5	2.2	2.0
15	15.0	10.0	7.5	6.0	5.0	4.3	3.7	3.3	3.0
20	20.0	13.3	10.0	8.0	6.7	5.7	5.0	4.4	4.0
25	25.0	16.7	12.5	10.0	8.3	7.1	6.2	5.6	5.0
30	30.0	20.0	15.0	12.0	10.0	8.6	7.5	6.7	6.0
35	35.0	23.3	17.5	14.0	11.7	10.0	8.7	7.8	7.0
40	40.0	26.7	20.0	16.0	13.3	11.4	10.0	8.9	8.0
45	45.0	30.0	22.5	18.0	15.0	12.8	11.2	10.0	9.0
50	50.0	33.3	25.0	20.0	16.7	14.3	12.5	11.1	10.0
55	55.0	36.7	27.5	22.0	18.3	15.7	13.7	12.2	11.0
60	60.0	40.0	30.0	24.0	20.0	17.1	15.0	13.3	12.0
65	65.0	43.3	32.5	26.0	21.7	18.6	16.2	14.4	13.0
70	70.0	46.7	35.0	28.0	23.3	20.0	17.5	15.6	14.0
75	75.0	50.0	37.5	30.0	25.0	21.4	18.7	16.7	15.0
80	80.0	53.3	40.0	32.0	26.7	22.8	20.0	17.8	16.0
85	85.0	56.7	42.5	34.0	28.3	24.3	21.2	18.9	17.0
90	90.0	60.0	45.0	36.0	30.0	25.7	22.5	20.0	18.0
95	95.0	63.3	47.5	38.0	31.7	27.1	23.7	21.1	19.0
100	100.0	66.7	50.0	40.0	33.3	28.6	25.0	22.2	20.0

[1]Values derived from the equation $Q = VA$, where Q is a peak runoff in cubic feet per second, V is velocity of flow in feet per second, and A is cross-section area of the ditch in square feet.

cross-section area of the ditch (3.2 ft^2) must remain the same. The depth of the ditch can be decreased and this will increase the width of the ditch. The result will be to increase the surface area of the ditch in contact with the flowing water and thereby decrease the water velocity. Reducing the gradient would also decrease velocity but may not be necessary in this case. If the depth is reduced to 1.0 foot, the width will be increased to 6.4 feet. The new hydraulic radius will be 0.453. The solution for the Manning equation under the new circumstance is:

$$V = \frac{1.49}{0.022} \times 0.589 \times 0.0707$$

$$V = 2.82 \text{ ft/s}$$

This velocity is close to the desired 2.5 ft/s and is probably acceptable. If the 2.82 ft/s is not acceptable, the depth can be decreased or the gradient can be decreased until the desired velocity is obtained.

Table 37. Widths and Depths Needed to Obtain Various Cross-Section Areas of Ditches[1]

Depth (ft)	Width (ft)									
	3	4	5	6	7	8	9	10	11	12
0.5	0.75	1.0	1.25	1.50	1.25	2.00	2.25	2.50	3.75	5.00
0.75	1.12	1.5	1.87	2.25	2.62	3.00	3.37	3.75	5.62	7.50
1.0	1.50	2.0	2.50	3.00	3.50	4.00	4.50	5.00	7.50	10.00
1.5	2.25	3.0	3.75	4.50	5.25	6.00	6.75	7.50	11.25	15.00
2.0	3.00	4.0	5.00	6.00	7.00	8.00	9.00	10.00	15.00	20.00
2.5	3.75	5.0	6.25	7.50	8.75	10.00	11.25	12.50	18.75	25.00
3.0	4.50	6.0	7.50	9.00	10.50	12.00	13.50	15.00	22.50	30.00
3.5	5.25	7.0	8.75	10.00	12.25	14.00	15.75	17.50	26.25	35.00
4.0	6.00	8.0	10.00	12.00	14.00	16.00	18.00	20.00	30.00	40.00
4.5	6.75	9.0	11.25	13.50	15.75	18.00	20.25	22.50	33.75	45.00
5.0	7.50	10.0	12.50	15.00	17.50	20.00	22.50	25.00	37.50	50.00

[1]Data given in square feet.

The depth under discussion is the depth of water in the ditch. The depth of the ditch should be increased by increasing the height of the constructed ridge (Figure 55) on the downhill side of the slope. This additional height is called freeboard and is the distance between the top of the ditch and the maximum water elevation expected. In diversions and terraces this freeboard is approximately 0.5 foot. The purpose of this freeboard is to safeguard against errors, soil settling, and unforeseen circumstances.

SIMPLIFIED DIVERSION DITCH DESIGN

The process of choosing the ditch width, depth, and gradient can be simplified by using prepared tables rather than solving the equations, as was done in the preceding section. The simplified method starts with the peak runoff expected for a given time period. The maximum channel velocity is then selected by using Table 32 or the reclamationist's experience. The peak runoff value and channel velocity can be used with Table 36 to determine the cross-section area of the channel that will be needed to carry the peak runoff. If the needed value does not occur in the table, it can be found by interpolation.

The cross-section area can then be used with Table 37 to find a suitable depth and width for the ditch. There can be several identical areas with different widths and depths. If a low velocity is desired, it is usually best to select a lesser depth value and greater width value in order to place more water in contact with the channel surface and thereby reduce velocity.

Table 38. Velocities[1] (ft/s) for Equilateral Triangular Cross-Section Channels Based on Hydraulic Radii (R) and Gradients (S)

R	S (ft/ft)												
(ft)	0.0025	0.005	0.010	0.015	0.020	0.025	0.030	0.035	0.040	0.045	0.050	0.055	0.060
0.20	1.16	1.64	2.32	2.84	3.28	3.66	4.01	4.33	4.63	4.91	5.18	5.43	5.67
0.25	1.34	1.90	2.69	3.29	3.80	4.25	4.66	5.03	5.38	5.70	6.01	6.30	6.58
0.30	1.52	2.15	3.04	3.72	4.29	4.80	5.26	5.68	6.07	6.44	6.79	7.12	7.43
0.35	1.68	2.38	3.36	4.12	4.76	5.32	5.83	6.29	6.73	7.14	7.52	7.89	8.24
0.40	1.84	2.60	3.68	4.50	5.20	5.81	6.37	6.88	7.35	7.80	8.22	8.62	9.01
0.45	1.99	2.81	3.98	4.87	5.62	6.29	6.89	7.44	7.95	8.44	8.89	9.33	9.74
0.50	2.13	3.02	4.27	5.23	6.03	6.75	7.39	7.98	8.53	9.05	9.54	10.01	10.45
0.55	2.27	3.21	4.55	5.57	6.43	7.19	7.87	8.51	9.09	9.64	10.17	10.66	11.14
0.60	2.41	3.41	4.82	5.90	6.81	7.62	8.34	9.01	9.64	10.22	10.77	11.30	11.80
0.65	2.54	3.59	5.08	6.22	7.19	8.04	8.80	9.51	10.16	10.78	11.36	11.92	12.45
0.70	2.67	3.78	5.34	6.54	7.55	8.44	9.25	9.99	10.68	11.33	11.94	12.52	13.00
0.75	2.79	3.95	5.59	6.85	7.91	8.84	9.68	10.46	11.18	11.86	12.50	13.11	13.69
0.80	2.92	4.13	5.84	7.15	8.25	9.23	10.11	10.92	11.67	12.38	13.05	13.69	14.30
0.90	3.16	4.46	6.31	7.73	8.93	9.98	10.53	11.37	12.15	12.89	13.59	14.25	14.89
1.00	3.39	4.79	6.77	8.29	8.58	10.71	11.73	12.67	13.55	14.37	15.14	15.88	16.59
1.05	3.50	4.95	7.00	8.57	9.89	11.06	12.12	13.09	13.99	14.84	15.65	16.41	17.14
1.10	3.61	5.10	7.22	8.84	10.21	11.41	12.50	13.50	14.43	15.31	16.14	16.93	17.68
1.15	3.72	5.26	7.43	9.10	10.51	11.75	12.88	13.91	14.87	15.77	16.62	17.43	18.21
1.20	3.82	5.41	7.65	9.37	10.82	12.09	13.25	14.31	15.30	16.22	17.10	17.94	18.73

1.25	3.93	5.56	7.86	9.63	11.11	12.43	13.61	14.70	15.72	16.67	17.57	18.43	19.25
1.30	4.03	5.70	8.07	9.88	11.41	12.76	13.97	15.09	16.13	17.11	18.04	18.92	19.76
1.35	4.14	5.85	8.27	10.13	11.70	13.08	14.33	15.48	16.55	17.55	18.50	19.40	20.26
1.40	4.24	5.99	8.48	10.38	11.99	13.40	14.68	15.86	16.95	17.98	18.95	19.88	20.76
1.45	4.34	6.14	8.68	10.63	12.27	13.72	15.03	16.23	17.35	18.41	19.40	20.35	21.25
1.50	4.44	6.28	8.87	10.87	12.55	14.03	15.37	16.60	17.75	18.83	19.84	20.81	21.74
1.55	4.54	6.41	9.07	11.11	12.83	14.34	15.71	16.97	18.14	19.24	20.28	21.27	22.22
1.60	4.63	6.55	9.27	11.35	13.10	14.65	16.05	17.33	18.53	19.65	20.72	21.73	22.69
1.65	4.73	6.69	9.46	11.58	13.37	14.95	16.38	17.69	18.91	20.06	21.15	22.18	23.17
1.70	4.82	6.82	9.65	11.82	13.64	15.25	16.71	18.05	19.19	20.46	21.57	22.62	23.63
1.75	4.92	6.95	9.84	12.05	13.91	15.55	17.04	18.40	19.67	20.86	21.99	23.07	24.09
1.80	5.01	7.09	10.02	12.27	14.17	15.85	17.36	18.75	20.04	21.26	22.41	23.50	24.55
1.85	5.10	7.22	10.21	12.50	14.43	16.14	17.68	19.10	20.41	21.65	22.82	23.94	25.00
1.90	5.20	7.35	10.39	12.72	14.69	16.43	18.00	19.44	20.78	22.04	23.23	24.37	25.45
1.95	5.29	7.48	10.57	12.95	14.95	16.71	18.31	19.78	21.14	22.43	23.64	24.79	25.89
2.00	5.38	7.60	10.75	13.17	15.20	17.00	18.62	20.11	21.50	22.81	14.04	25.21	26.34
2.05	5.47	7.73	10.93	13.39	15.46	17.28	18.83	20.45	21.86	23.18	24.44	25.63	26.77
2.10	5.55	7.85	11.11	13.60	15.71	17.56	19.24	20.78	22.21	23.56	24.84	26.05	27.21
2.15	5.64	7.98	11.28	13.82	15.96	17.84	19.54	21.11	22.56	23.93	25.23	26.46	27.64
2.20	5.73	8.10	11.46	14.03	16.20	18.11	19.84	21.43	22.91	24.30	25.62	26.87	28.06

[1]These values were obtained by solving Manning's equation with an n value of 0.022.

The selected depth and width can now be used with Table 33 to obtain a value for the hydraulic radius (R) of the channel. The R value is then used with Table 38 to find the gradient that will give the desired velocity. If the desired velocity cannot be obtained with the R value, a new depth and width that will produce a smaller R value must be selected and the process repeated.

Example: Assume that the peak runoff (Q) is 30 ft^3/s and the desired velocity is 3.0 ft/s. Table 36 shows that the cross-section area for a ditch that would carry this volume and velocity is 10 ft^2. Table 37 shows that a triangular ditch that is 2.5 ft deep and 8 ft wide will have a cross section of 10 ft^2. Table 33 shows an R value of 1.060 for this ditch. Table 38 shows a velocity of 3.50 ft/s at the smallest S value for the R value of 1.05. Therefore, an R value of 1.06 is unacceptable and a new depth and width must be chosen. The new procedure would be:

Table 37: 2 ft deep and 10 ft wide

Table 33: $R = 0.928$ ft

Table 38: $V = 3.20$ ft/s at a gradient of 0.0025 ft/ft

This value is close to the desired velocity; however, the gradient is small and the reclamationist may prefer a larger gradient. In this case, return to Table 37 and select a lesser depth that will yield 10 ft^2 or an approximation:

Table 37: 1 ft deep and 12 ft wide

Table 33: $R = 0.493$ ft

Table 38: $V = 3.02$ ft/s at a gradient of 0.005 ft/ft

Now the velocity and gradient are acceptable, but the ditch is shallow and wide. If the ditch dimensions are unacceptable, the reclamationist will have to accept a greater velocity in order to reduce the ditch width or increase the depth.

TRAPEZOIDAL CHANNEL DESIGN

Another useful cross-section shape for diversion ditches is the trapezoid. This shape is convenient to build and keeps much of the water in contact with the ditch surface to reduce velocity. Trapezoidal ditches can be designed by using the same two equations that were used to design triangular ditches. Figure 61 details how to calculate ditch cross-section values to be used in Manning's equation. However, a faster and simpler system of designing a trapezoidal system has been devised.

The starting point is the same as for a triangular ditch. The peak runoff and then the maximum flow velocity that is desirable must be estimated (Table 32). The peak runoff and maximum velocity can then be used to determine the needed cross-section area for the trapezoidal

Table 39. Side Slope, Bottom Width, Depth, Top Width, Area, and Hydraulic Radius for Trapezoidal Diversion Ditches of Varying Design

Design no.	Side slope	Bottom width (*B*) (ft)	Depth (*D*) (ft)	Top width (*T*) (ft)	Cross-section area (*A*) (ft^2)	Hydraulic radius (*R*) (ft)
1	1.5:1	2	0.5	3.5	1.38	0.362
2	2:1	2	0.5	4.0	1.50	0.354
3	3:1	2	0.5	5.0	1.75	0.339
4	1.5:1	2	1.0	5.0	3.50	0.525
5	2:1	2	1.0	6.0	4.00	0.617
6	3:1	2	1.0	8.0	5.00	0.601
7	1.5:1	3	0.5	4.5	1.97	0.391
8	2:1	3	0.5	5.0	2.00	0.382
9	3:1	3	0.5	6.0	2.25	0.365
10	1.5:1	3	1.0	6.0	4.50	0.682
11	2:1	3	1.0	7.0	5.00	0.668
12	3:1	3	1.0	9.0	6.00	0.644
13	1.5:1	3	1.5	7.5	7.88	0.938
14	2:1	3	1.5	9.0	9.00	0.928
15	3:1	3	1.5	12.0	11.25	0.901
16	1.5:1	3	2.0	9.0	12.00	1.174
17	2:1	3	2.0	11.0	14.00	1.172
18	3:1	3	2.0	15.0	18.00	1.151
19	1.5:1	4	0.5	5.5	2.37	0.409
20	2:1	4	0.5	6.0	2.50	0.401
21	3:1	4	0.5	7.0	2.75	0.384
22	1.5:1	4	1.0	7.5	5.50	0.724
23	2:1	4	1.0	8.0	6.00	0.707
24	3:1	4	1.0	10.0	7.00	0.678
25	1.5:1	4	1.5	8.5	9.37	0.997
26	2:1	4	1.5	10.0	10.50	0.979
27	3:1	4	1.5	13.0	12.75	0.946
28	1.5:1	4	2.0	10.0	14.00	1.250
29	2:1	4	2.0	12.0	16.00	1.235
30	3:1	4	2.0	16.0	20.00	1.202
31	1.5:1	4	3.0	13.0	25.50	1.721
32	2:1	4	3.0	16.0	30.00	1.722
33	3:1	4	3.0	22.0	39.00	1.697
34	1.5:1	5	0.5	6.5	3.25	0.423
35	2:1	5	0.5	7.0	3.50	0.415
36	3:1	5	0.5	8.0	4.00	0.398
37	1.5:1	5	1.0	8.0	6.50	0.755
38	2:1	5	1.0	9.0	7.00	0.739
39	3:1	5	1.0	11.0	8.00	0.708

continued

Table 39. Side Slope, Bottom Width, Depth, Top Width, Area, and Hydraulic Radius for Trapezoidal Diversion Ditches of Varying Design *(Continued)*

Design no.	Side slope	Bottom width (*B*) (ft)	Depth (*D*) (ft)	Top width (*T*) (ft)	Cross-section area (*A*) (ft^2)	Hydraulic radius (*R*) (ft)
40	1.5:1	5	3.0	14.0	28.50	1.802
41	2:1	5	3.0	17.0	33.00	1.792
42	3:1	5	3.0	23.0	42.00	1.751
43	1.5:1	6	0.5	7.5	3.37	0.433
44	2:1	6	0.5	8.0	3.50	0.425
45	3:1	6	0.5	9.0	3.75	0.304
46	1.5:1	6	1.0	9.0	7.50	0.781
47	2:1	6	1.0	10.0	8.00	0.763
48	3:1	6	1.0	12.0	9.00	0.730
49	1.5:1	6	1.5	10.5	12.37	1.085
50	2:1	6	1.5	12.0	13.50	1.061
51	3:1	6	1.5	15.0	15.75	1.071
52	1.5:1	6	2.0	12.0	18.00	1.364
53	2:1	6	2.0	14.0	20.00	1.337
54	3:1	6	2.0	18.0	24.00	1.288
55	1.5:1	6	3.0	15.0	31.50	1.875
56	1.5:1	6	3.0	18.0	36.00	1.852
57	3:1	6	4.0	24.0	45.00	1.803
58	1.5:1	6	4.0	18.0	48.00	2.351
59	2:1	6	4.0	22.0	56.00	2.345
60	3:1	6	4.0	30.0	72.00	2.300
61	1.5:1	7	0.5	8.5	4.25	0.480
62	2:1	7	0.5	9.0	4.50	0.433
63	3:1	7	0.5	10.0	5.00	0.418
64	1.5:1	7	1.0	10.0	8.50	0.801
65	2:1	7	1.0	11.0	9.00	0.784
66	3:1	7	1.0	13.0	10.00	0.750
67	1.5:1	7	1.5	11.5	12.75	1.118
68	2:1	7	1.5	13.0	13.50	1.094
69	3:1	7	1.5	16.0	15.00	1.046
70	1.5:1	8	0.5	9.5	4.37	0.446
71	2:1	8	0.5	10.0	4.50	0.439
72	3:1	8	0.5	11.0	4.75	0.426
73	1.5:1	8	1.0	11.0	9.50	0.819
74	3:1	8	1.0	12.0	10.00	0.801
75	3:1	8	1.0	14.0	11.00	0.768
76	1.5:1	8	1.5	12.5	15.37	1.147
77	2:1	8	1.5	14.0	16.50	1.121
78	3:1	8	1.5	17.0	18.75	1.073

continued

Table 39. *(Continued)*

Design no.	Side slope	Bottom width (*B*) (ft)	Depth (*D*) (ft)	Top width (*T*) (ft)	Cross section area (*A*) (ft^2)	Hydraulic radius (*R*) (ft)
79	1.5:1	8	2.0	14.0	19.00	1.446
80	2:1	8	2.0	16.0	20.00	1.416
81	3:1	8	2.0	20.0	22.00	1.356
82	1.5:1	8	2.5	15.5	23.75	1.727
83	2:1	8	2.5	18.0	25.00	1.694
84	3:1	8	2.5	23.0	27.50	1.627
85	1.5:1	8	3.0	17.0	28.50	1.993
86	2:1	8	3.0	20.0	30.00	1.961
87	3:1	8	3.0	26.0	33.00	1.891

ditch (Table 36). The next step is to proceed to Table 39 and use the indicated area to choose the side slope, bottom width, top width, and depth for the trapezoidal ditch that will provide the desired cross-section area. The final step is to take the *R* value from Table 39 that corresponds to the area, and then proceed to Table 40. This table can now be used to determine the channel gradient that will be necessary to obtain the desired velocity. If the desired velocity cannot be found in the row with the original *R* value, then return to Table 39 and select another channel design and *R* value that have the desired area or approximate desired area. Take this new *R* value and go back to Table 40 to find the desired velocity. All the factors, top width, bottom width, depth, side slope, and channel gradient are now known for the construction of a successful trapezoidal diversion ditch.

Example: The peak runoff is predicted to be 30 ft^3/s and the area needed to carry this flow at the desired velocity of 3.0 ft/s is 10.0 ft^2 (Table 36). Design no. 66 in Table 39 gives a cross-section area of 10.0 ft^2. Table 40 shows that the least velocity for design no. 66 is 2.80 ft/s at a gradient of 0.0025 ft/ft. This is close enough to the desired velocity. There are other designs that will produce a velocity near the desired velocity and one of these designs may be easier to construct. For instance, design no. 26 has an area of 10.5 ft^2 and the velocity is 3.34 ft/s at a gradient of 0.0025 ft/ft. Also, design no. 15 has an area of 11.25 ft^2 and the velocity would be 3.16 ft/s at a gradient of 0.0025 ft/ft. One of these designs might be preferable to design no. 66 because of equipment or terrain restrictions.

Table 40. Velocities[1] (ft/s) Obtained with Varying Hydraulic Radii (*R*) and Gradients (*S*) for Trapezoidal Cross-Section Channels

Design no.	*R* (ft)	*S* (ft/ft) 0.0025	0.005	0.010	0.015	0.020	0.025	0.030	0.035	0.040	0.045	0.050	0.055	0.060	0.065
1	0.362	1.72	2.43	3.44	4.21	4.86	5.44	5.96	6.44	6.88	7.30	7.69	8.07	8.43	8.77
2	0.354	1.69	2.40	3.39	4.15	4.79	5.36	5.87	6.34	6.78	7.19	7.58	7.95	8.30	8.64
3	0.339	1.64	2.33	3.29	4.03	4.66	5.21	5.70	6.16	6.59	6.98	7.36	7.72	8.07	8.39
4	0.625	2.48	3.58	4.95	6.86	7.00	7.83	8.58	9.26	9.90	10.50	11.07	11.61	12.13	12.62
5	0.617	2.45	3.47	4.91	6.01	6.94	7.76	8.50	9.18	9.82	10.41	10.98	11.51	12.02	12.51
6	0.601	2.41	3.41	4.82	5.91	6.82	7.63	8.35	9.02	9.65	10.23	10.79	11.31	11.81	12.30
7	0.391	1.81	2.56	3.62	4.44	5.12	5.73	6.27	6.77	7.24	7.68	8.10	8.49	8.87	9.23
8	0.382	1.78	2.52	3.57	4.37	5.04	5.64	6.18	6.67	7.13	7.56	7.97	8.36	8.73	9.09
9	0.365	1.73	2.45	3.46	4.24	4.89	5.47	5.99	6.47	6.92	7.34	7.73	8.11	8.47	8.82
10	0.682	2.62	3.71	5.25	6.43	7.42	8.30	9.09	9.82	10.49	11.13	11.73	12.31	12.85	13.38
11	0.668	2.59	3.66	5.18	6.34	7.32	8.18	8.96	9.68	10.35	10.98	11.57	12.14	12.68	13.19
12	0.644	2.53	3.57	5.05	6.19	7.14	7.99	8.75	9.45	10.10	10.71	11.29	11.84	12.37	12.88
13	0.938	3.24	4.59	6.49	7.95	9.18	10.26	11.24	12.14	12.98	13.77	14.51	15.22	15.90	16.55
14	0.928	3.22	4.56	6.44	7.89	9.11	10.19	11.16	12.05	12.89	13.67	14.41	15.11	15.78	16.43
15	0.901	3.16	4.47	6.32	7.74	8.93	9.99	10.94	11.82	12.64	13.40	14.13	14.82	15.48	16.11
16	1.174	3.77	5.33	7.54	9.23	10.66	11.92	13.05	14.10	15.07	15.99	16.85	17.68	18.46	19.22
17	1.174	3.76	5.32	7.53	9.22	10.65	11.90	13.04	14.08	15.06	15.97	16.83	17.66	18.44	19.19
18	1.151	3.72	5.26	7.44	9.11	10.52	11.76	12.88	13.92	14.88	15.78	16.63	17.44	18.22	18.96
19	0.409	1.87	2.64	3.73	4.57	5.28	5.90	6.46	6.98	7.46	7.92	8.34	8.75	9.14	9.51
20	0.401	1.84	2.60	3.68	4.51	5.21	5.82	6.38	6.89	7.37	7.81	8.24	8.64	9.02	9.39

21	0.384	1.79	2.53	3.58	4.38	5.06	5.66	6.20	6.69	7.16	7.59	8.00	8.39	8.76	9.12
22	0.724	2.73	3.86	5.46	6.69	7.72	8.63	9.46	10.22	10.92	11.58	12.21	12.81	13.38	13.92
23	0.707	2.69	3.80	5.37	6.58	7.60	8.50	9.31	10.06	10.75	11.40	12.02	12.61	13.17	13.70
24	0.678	2.61	3.70	5.23	6.40	7.39	8.26	9.05	9.78	10.45	11.09	11.69	12.26	12.80	13.33
25	0.997	3.38	4.78	6.76	8.28	9.56	10.69	11.71	12.65	13.52	14.34	15.11	15.85	16.56	17.23
26	0.979	3.34	4.72	6.68	8.18	9.44	10.56	11.57	12.49	13.36	14.17	14.93	15.66	16.36	17.02
27	0.946	3.26	4.62	6.53	7.99	9.23	10.32	11.30	12.21	13.05	13.85	14.59	15.31	15.99	16.64
28	1.250	3.93	5.56	7.86	9.63	11.11	12.43	13.61	14.70	15.72	16.67	17.57	18.43	19.25	20.04
29	1.235	3.90	5.51	7.80	9.55	11.03	12.33	13.50	14.59	15.59	16.54	17.43	18.28	19.10	19.88
30	1.202	3.83	5.41	7.66	9.38	10.83	12.11	13.26	14.32	15.31	16.24	17.12	17.96	18.75	19.52
31	1.721	4.86	6.88	9.73	11.91	13.76	15.38	16.85	18.20	19.45	20.63	21.75	22.81	23.83	24.80
32	1.722	4.87	6.88	9.73	11.92	13.76	15.38	16.85	18.20	19.46	20.64	21.76	22.82	23.83	24.81
33	1.697	4.82	6.81	9.64	11.80	13.63	15.24	16.69	18.03	19.27	20.44	21.55	22.60	23.60	24.57
34	0.423	1.91	1.70	3.82	4.67	5.40	6.03	6.61	7.14	7.63	3.09	8.65	8.95	9.35	9.73
35	0.415	1.88	2.66	3.77	4.61	5.33	5.96	6.52	7.05	7.53	7.99	8.42	8.83	9.23	9.60
36	0.415	1.83	2.59	3.66	4.49	5.18	5.79	6.35	6.85	7.33	7.77	8.19	8.59	8.97	9.34
37	0.755	2.81	3.97	5.62	6.88	7.94	8.88	9.73	10.50	11.23	11.91	12.56	13.17		
38	0.739	2.77	3.91	5.54	6.78	7.83	8.75	9.59	10.36	11.38	11.74	12.38	12.98		
39	0.706	2.68	3.80	5.37	6.58	7.59	8.49	9.30	10.04	10.74	11.39	12.01	12.59		
40	1.802	5.01	7.09	10.03	12.28	14.18	15.86	17.37	18.76	20.06	21.28	22.43	23.52	24.57	25.57
41	1.792	5.00	7.07	9.99	12.24	14.13	15.80	17.31	18.69	19.98	21.20	22.34	23.43	24.48	25.48
42	1.751	4,92	6.96	9.84	12.05	13.91	15.56	17.04	18.41	19.68	21.87	22.00	23.08	24.10	25.09
43	0.433	1.94	2.74	3.88	4.75	5.48	6.13	6.71	7.25	7.75	8.22	8.67	9.09	9.49	9.88
44	0.425	1.91	2.71	3.83	4.69	5.41	6.05	6.63	7.16	7.66	8.12	8.56	8.98	9.38	9.76
45	0.304	1.53	2.17	3.06	3.75	4.33	4.84	5.30	5.73	6.12	6.50	6.85	7.18	7.50	7.81
46	0.781	2.87	4.06	5.74	7.03	8.12	9.08	9.95	10.75	11.49	12.18	12.84	13.47	14.07	14.64
47	0.763	2.83	4.00	5.66	6.93	8.00	8.94	9.79	10.58	11.31	12.00	12.65	13.26	13.85	14.42
48	0.730	2.75	3.88	5.49	6.72	7.77	8.68	9.51	10.27	10.98	11.65	12.28	12.88	13.45	14.00
49	1.085	3.58	5.06	7.15	8.76	10.11	11.31	12.39	13.38	14.30	15.17	15.99	16.77	17.52	18.23
50	1.061	3.52	4.98	7.05	8.63	9.96	11.14	12.20	13.18	14.09	14.95	15.75	16.52	17.26	17.96

continued

Table 40. Velocities[1] (ft/s) Obtained with Varying Hydraulic Radii (R) and Gradients (S) for Trapezoidal Cross-Section Channels *(Continued)*

Design no.	R (ft)	S (ft/ft) 0.0025	0.005	0.010	0.015	0.020	0.025	0.030	0.035	0.040	0.045	0.050	0.055	0.060	0.065
51	1.071	3.54	5.01	7.09	8.68	10.03	11.21	12.28	13.26	14.18	15.04	15.85	16.63	17.37	18.08
52	1.364	4.16	5.89	8.33	10.20	11.78	13.17	14.43	15.58	16.66	17.67	18.63	19.54	20.40	21.24
53	1.337	4.11	5.81	8.22	10.07	11.62	13.00	14.24	15.38	16.44	17.44	18.38	19.28	20.13	20.96
54	1.288	4.01	5.67	8.02	9.82	11.34	12.68	13.89	15.00	16.04	17.01	17.93	18.80	19.64	20.44
55	1.875	5.15	7.28	10.30	12.61	14.56	16.28	17.84	19.27	20.60	21.85	23.03	24.15	25.23	26.26
56	1.852	5.11	7.22	10.21	12.51	14.44	16.15	17.69	19.11	20.43	21.67	22.84	23.95	25.02	26.04
57	1.803	5.02	7.09	10.03	12.29	14.19	15.86	17.38	18.77	20.07	21.28	22.43	23.53	24.58	25.58
58	2.351	5.99	8.47	11.98	14.67	16.94	18.93	20.74	22.40	23.95	25.40	26.78	28.08	29.33	30.53
59	2.345	5.98	8.45	11.95	14.64	16.91	18.90	20.71	22.37	23.91	25.38	26.73	28.04	29.28	30.48
60	2.300	5.90	8.34	11.80	14.45	16.69	18.66	20.44	22.08	23.60	25.03	26.39	27.68	28.91	30.09
61	0.440	1.96	2.77	3.92	4.80	5.54	6.19	6.78	7.33	7.83	8.31	8.76	9.19	9.59	9.99
62	0.433	1.94	2.74	3.88	4.75	5.48	8.13	6.71	7.25	7.75	8.22	8.67	9.09	9.49	9.88
63	0.418	1.89	2.68	3.79	4.64	5.35	5.98	6.58	7.08	7.57	8.03	8.46	8.88	9.27	9.65
64	0.801	2.92	4.13	5.84	7.15	8.26	9.24	10.12	10.93	11.68	12.39	13.06	13.70		
65	0.784	2.88	4.07	5.78	7.05	8.14	9.10	9.97	10.77	11.52					
66	0.750	2.80	3.95	5.59	6.85	7.91	8.84	9.68	10.46	11.18					
67	1.118	3.65	5.16	7.30	8.94	10.32	11.54	12.64	13.65	14.59					
68	1.094	3.60	5.08	7.19	8.81	10.17	11.37	12.46	13.45	14.38					
69	1.046	3.49	4.93	6.98	8.55	9.87	11.03	12.09	13.06	13.96					
70	0.446	1.97	2.80	3.95	4.84	5.59	6.25	6.85	7.40	7.91	8.39	8.84	9.27	9.68	10.08

71	0.439	1.96	2.77	3.91	4.79	5.53	6.19	6.78	7.32	7.82	8.30	8.75	9.17	9.58	9.97
72	0.426	1.92	2.71	3.83	4.70	5.42	6.06	6.64	7.17	7.67	8.13	8.57	8.99	9.39	9.78
73	0.819	2.96	4.19	5.93	7.26	8.38	9.37	10.27	11.09	11.86	12.58	13.26	13.90	14.52	15.11
74	0.801	2.92	4.13	5.84	7.15	8.26	9.24	10.12	10.93	11.68	12.39	13.06	13.70	14.31	14.89
75	0.768	2.34	4.02	5.68	6.96	8.03	8.98	9.84	10.63	11.36	12.05	12.70	13.32	13.91	14.48
76	1.147	3.71	5.25	7.42	9.09	10.50	11.73	12.85	13.88	14.84	15.74	16.59	17.40	18.18	18.92
77	1.121	3.35	5.17	7.31	8.95	10.34	11.56	12.66	13.67	14.62	15.50	16.34	17.14	17.90	18.63
78	1.073	3.55	5.02	7.10	8.69	10.04	11.22	12.29	13.28	14.20	15.06	15.87	16.65	17.39	18.10
79	1.446	4.33	6.12	8.66	10.61	12.25	13.70	15.00	16.20	17.32	18.37	19.37	20.31	21.22	22.08
80	1.416	4.27	6.04	8.54	10.46	12.08	13.51	14.79	15.98	17.08	18.12	19.10	20.03	20.92	21.78
81	1.356	4.15	5.87	8.30	10.16	11.74	13.12	14.37	15.52	16.60	17.60	18.56	19.46	20.33	21.16
82	1.727	4.88	6.89	9.75	11.94	13.79	15.42	16.89	18.24	19.50	20.69	21.80	22.87	23.89	24.86
83	1.694	4.81	6.81	9.63	11.79	13.61	15.22	16.67	18.01	19.25	20.42	21.53	22.58	23.58	24.54
84	1.627	4.69	6.63	9.37	11.48	13.25	24.82	16.23	17.53	18.74					
85	1.993	5.36	7.58	10.37	13.14	15.17	16.96	18.58	20.07	21.46					
86	1.961	5.31	7.50	10.61	13.00	15.01	16.78	18.38	19.86	21.23					
87	1.891	5.18	7.33	10.36	12.69	14.65	16.38	17.94	19.38	20.72					

[1]These values were obtained by solving Manning's equation with an n value of 0.022.

Table 41. Top Width, Depth, Area, and Hydraulic Radius for Parabolic Diversion Ditches of Varying Design

Design no.	Width (*T*) (ft)	Depth (*D*) (ft)	Cross-section area (*A*) (ft^2)	Hydraulic radius (*R*) (ft)
1	4	0.5	1.33	0.362
2	4	1.0	2.67	0.571
3	4	1.5	4.00	0.727
4	6	0.5	2.00	0.327
5	6	1.0	4.00	0.621
6	6	1.5	6.00	0.857
7	6	2.0	8.00	1.029
8	8	0.5	2.67	0.330
9	8	1.0	5.34	0.640
10	8	1.5	8.00	0.914
11	8	2.0	10.67	1.143
12	8	2.5	13.34	1.322
13	8	3.0	16.01	1.455
14	10	0.5	3.33	0.331
15	10	1.5	10.00	0.943
16	10	1.7	11.33	1.053
17	10	2.0	13.32	1.205
18	10	2.5	16.67	1.429
19	10	3.0	20.01	1.613
20	12	0.5	4.00	0.332
21	12	1.0	8.00	0.960
22	12	1.5	12.00	1.241
23	12	2.0	16.00	1.484
24	12	2.5	20.00	1.494
25	12	2.5	24.00	1.714
26	12	3.0	28.00	1.902
27	14	0.5	4.67	0.332
28	14	1.0	9.34	0.658
29	14	1.5	14.01	0.970
30	14	2.0	18.68	1.265
31	14	2.5	23.35	1.536
32	14	3.0	28.02	1.782
33	14	3.5	32.69	2.000
34	14	4.0	37.36	2.190
35	16	0.5	5.35	0.332
36	16	1.0	10.70	0.660
37	16	1.5	16.01	0.977
38	16	2.0	21.36	1.280
39	16	2.5	26.71	1.565
40	16	3.0	32.06	1.629

continued

Table 41. *(Continued)*

Design no.	Width (T) (ft)	Depth (D) (ft)	Cross section area (A) (ft^2)	Hydraulic radius (R) (ft)
41	16	3.5	37.41	2.069
42	16	4.0	42.69	2.286
43	18	1.0	12.00	0.661
44	18	2.0	24.00	1.291
45	18	3.0	36.00	1.862
46	18	4.0	48.00	2.356
47	18	5.0	60.00	2.765
48	20	2.0	36.68	1.299
49	20	4.0	53.36	2.410
50	20	6.0	80.04	3.226
51	22	2.0	29.35	1.305
52	22	4.0	58.70	2.451
53	22	6.0	88.04	3.338
54	24	4.0	64.03	2.483
55	24	6.0	96.05	3.429
56	24	8.0	128.06	4.114

PARABOLIC CHANNEL DESIGN

The third useful cross-section shape for diversion ditches is the parabolic shape. This shape lends itself to producing a slow water velocity. Also it is likely that no matter what the original shape may be, it will tend to be parabolic after a period of time if erosive water velocities are allowed.

To design a parabolic ditch, the peak runoff must be determined as for the other ditch shapes. Next the maximum permissible velocity must be chosen (Table 32). The peak runoff and maximum velocity values can be applied in Table 36 to obtain the needed cross-section area for the ditch to carry the peak runoff at the desired velocity. This cross-section area can be used with Table 41 to determine the depth and top width needed for a parabolic ditch of the needed cross-section area. This table also shows the hydraulic radius for such a ditch. This R value can now be used in Table 42 to select a channel gradient (S) that will give the desired water velocity or a value close to the desired velocity. If the desired velocity is not found for these R and S values, Table 41 must be referred to again and a new width and depth chosen that will give a cross-section area close to the desired area. The R value for this new width and depth can then be applied in Table 42 to determine the needed gradient. These operations can be repeated until an acceptable depth, width, and gradient are found that will provide the desired water velocity in the parabolic ditch channel.

Table 42. Velocities[1] (ft/s) Obtained with Varying Hydraulic Radii (*R*) and Gradients (*S*) for Parabolic Cross-Section Channels

Design no.	*R* (ft)	*S* (ft/ft) 0.0025	0.005	0.010	0.015	0.020	0.025	0.030	0.035	0.040	0.045	0.050	0.055	0.060	0.065
1	0.320	1.58	2.24	3.17	3.88	4.48	5.01	5.49	5.93	6.34	6.72	7.08	7.43	7.76	8.08
2	0.571	2.33	3.30	4.66	5.71	6.59	7.37	8.07	8.72	9.32	9.89	10.42	10.93	11.42	11.88
3	0.727	2.74	3.87	5.48	6.71	7.74	8.66	9.48	10.24	10.95	11.62	12.24	12.84	13.41	13.96
4	0.327	1.61	2.27	3.21	3.94	4.55	5.08	5.57	6.01	6.43	6.82	7.19	7.54	7.87	8.20
5	0.621	2.46	3.49	4.93	6.04	6.97	7.79	8.54	9.22	9.86	10.46				
6	0.857	3.06	4.32	6.11	7.48	8.64	9.66	10.58	11.43						
7	1.029	3.45	4.88	6.90	8.45	9.76	10.91	11.96							
8	0.330	1.62	2.29	3.23	3.96	4.57	5.11	5.60	6.05	6.47	6.86	7.23	7.58	7.92	8.26
9	0.640	2.51	3.56	5.03	6.16	7.11	7.95	8.71	9.41	10.06	10.67	11.25			
10	0.914	3.19	4.51	6.38	7.81	9.02	10.47	11.71							
11	1.143	3.70	5.24	7.40	9.07	10.47	11.71								
12	1.322	4.00	5.77	8.16	9.99	11.54									
13	1.455	4.35	6.15	8.70	10.65										
14	0.331	1.62	2.29	3.24	3.97	4.58	5.12	5.61	6.06	6.48	6.87	7.24	7.60	7.94	8.26
15	0.649	2.54	3.59	5.08	6.22	7.18	8.03	8.79	9.50	10.15	10.77	11.35			
16	0.843	3.26	4.61	6.51	7.98	9.21	10.30	11.28							
17	1.205	3.83	5.42	7.67	9.39	10.85	12.13								
18	1.429	4.30	6.08	8.59	10.52	12.15									
19	1.613	4.66	6.59	9.32	11.41										
20	0.332	1.62	2.30	3.25	3.98	4.59	5.13	5.62	6.07	6.49	6.89	7.26	7.62	7.95	8.28

21	0.655	2.55	3.61	5.11	6.26	7.22	8.08	8.85	9.58	10.22	10.84	11.42			
22	0.960	3.30	4.68	6.59	8.07	9.32	10.42	11.42							
23	1.241	3.91	5.53	7.82	9.58	11.08									
24	1.494	4.43	6.26	8.85	10.84										
25	1.714	4.85	6.86	9.70	11.88										
26	1.902	5.02	7.35	10.40											
27	0.332	1.62	2.30	3.25	3.98	4.59	5.13	5.62	6.07	6.49	6.89	7.26	7.62	7.95	8.28
28	0.658	2.56	3.62	5.12	6.28	7.25	8.10	8.87	9.59	10.25	10.87	11.46			
29	0.970	3.32	4.69	6.64	8.13	9.39	10.49	11.49							
30	1.265	3.96	5.60	7.92	9.70	11.20									
31	1.536	4.51	6.38	9.02	11.04										
32	1.782	4.98	7.04	9.96	12.19										
33	2.000	5.38	7.60	10.75											
34	2.190	5.71	8.08	11.42											
35	0.332	1.62	2.30	3.25	3.98	4.59	5.13	5.62	6.07	6.49	6.89	7.26	7.62	7.95	8.28
36	0.660	2.57	3.63	5.13	6.29	7.26	8.12	8.89	9.60	10.27	10.89				
37	0.977	3.33	4.72	6.67	8.17	9.43	10.54	11.55							
38	1.280	3.99	5.65	7.98	9.78	11.29									
39	1.565	4.56	6.46	9.13	11.18										
40	1.629														
41	2.069	5.50	7.78	11.00	13.47										
42	2.286	5.88	8.31	11.75	14.59										
43	0.661	2.57	3.63	5.14	6.29	7.27	8.13	8.90	9.61	10.28	10.90	11.49			
44	1.291	4.02	5.68	8.03	9.83	11.36	12.70								
45	1.862	5.13	7.25	10.25	12.55										
46	2.356	6.00	8.48	11.99	14.69										
47	2.765	6.67	9.43	13.34											
48	1.299	4.03	5.70	8.06	9.88	11.40	12.75								
49	2.410	6.09	8.61	12.17	14.91										
50	3.226	7.39	10.46	14.79											

continued

Table 42. Velocities[1] (ft/s) Obtained with Varying Hydraulic Radii (R) and Gradients (S) for Parabolic Cross-Section Channels *(Continued)*

Design no.	R (ft)	S (ft/ft) 0.0025	0.005	0.010	0.015	0.020	0.025	0.030	0.035	0.040	0.045	0.050	0.055	0.060	0.065
51	1.305	4.04	5.72	8.09	9.91	11.44	12.79								
52	2.451	6.16	8.71	12.31	15.08										
53	3.338	7.56	10.70	15.13											
54	2.483	6.21	8.78	12.42											
55	3.429	7.70	10.89	15.40											
56	4.114	8.69	12.30	17.39											

[1]These values were obtained by solving Manning's equation with an n value of 0.022.

DIVERSION DITCH OUTLETS

The water that is carried by diversion ditches or terraces must be moved from the end of the diversion or terrace to a sediment pond or natural stream. The structure that moves this water is called an outlet or waterway. These outlets are usually designed to accept all the water from the diversions without eroding significantly. This can be done by constructing the outlets wide enough to dissipate the water from the diversions and reduce the speed of flow. Another method of preventing erosion in the outlet is to line the outlet channel with an erosion resistant material such as vegetation, stone, concrete, or one of the erosion control materials manufactured especially for this purpose.

It is possible that the reclamationist will decide simply to allow the ditches to empty into some convenient area without constructing a ditch outlet. This may be especially true if there is a thin cover of topsoil with noneroding, stony overburden beneath the topsoil. It may be less expensive and time-consuming to accept the localized erosion in the outlet and repair the damage after vegetation becomes established between the diversion ditches. This vegetation between diversion ditches will lessen the amount of water flowing through the ditches and the outlet, and a reduction in water flow in the outlet may allow the repair and revegetation of the outlet at a later date.

When locating outlets in undisturbed areas, they can be placed in natural depressions or hillside folds and the diversion ditches emptied into them. However, because reclamation is a continuous operation, the outlets must be placed at the ends of the diversion ditches wherever the reclamation operation stops for the season. This means that outlets are often placed in inconvenient locations for discharge into sediment ponds.

It takes weeks or months to establish vegetation in the bed of an outlet. Therefore stone or some other material is often used to prevent erosion of the outlet channel when immediate protection is needed. Stone is usually one of the cheapest and most available materials for lining an outlet on a surface mine. This stone lining is called riprap. Riprap should be a mixture of different size stones. An ideal riprap mixture would grade uniformly downward in size from the largest sizes to a minimum size of no less than 1 inch. The smaller stones are needed to fill the voids between the larger stones in order to prevent the eroding of soil material from beneath the riprap. Riprap in graduated sizes may not be needed if one of the woven flexible materials is placed on the soil surface beneath the riprap.

The first piece of information needed to design an outlet is the maximum amount of water that the outlet will have to carry from the diversion ditches to the sediment pond or natural body of water. This amount of runoff can be calculated by using the Beasley equation, the Soil Conservation Service method, or the rational formula as described in the runoff section.

After the peak flow is determined, the channel of the outlet can be designed. The simplest design is based solely on the peak flow and a

Table 43. Bottom Widths, Channel Depths, and Side Slopes for Trapezoidal[1] Diversion Ditch Outlets Designed to Carry Various Peak Flows

Q (ft^3/s)	Bottom width (ft)	Channel depth (ft)	Side slope (x:1)	Min. ave. stone diam[2] (in)	Largest stone diam[3] (in)	Thickness of riprap layer[4] (in)
0-30	4	1.5	3.0:1	5.0	7.50	11.25
30-50	6	1.5	3.0:1	5.5	8.25	12.37
50-65	8	1.5	3.0:1	6.0	9.00	13.50
65-80	10	1.5	3.0:1	7.0	10.50	15.75
80-100	12	1.5	3.0:1	7.5	11.25	16.87
100-200	12	1.5	3.0:1	10.5	15.75	23.62
200-300	13	2.0	2.5:1	11.5	16.50	24.75
300-400	17	2.0	2.5:1	12.5	18.75	28.12
400-500	20	2.5	2.5:1	14.0	21.00	31.50
500-600	20	2.5	2.5:1	15.5	23.25	34.87
600-700	22	2.5	2.5:1	16.0	24.00	36.00
700-800	22	2.5	2.5:1	17.0	25.50	38.25
800-900	22	3.0	2.5:1	18.0	27.00	40.50
900-1000	22	3.0	2.5:1	19.5	29.25	43.87

[1]A maximum channel slope of 30 percent was assumed.

[2]Minimum average stone diameter: 50 percent of the stone, by weight, smaller than this value.

[3]1.5 times the minimum average stone diameter.

[4]1.5 times the largest stone diameter.

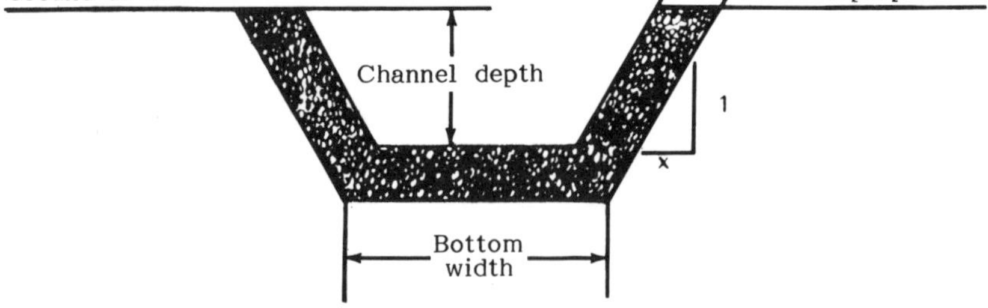

maximum channel slope. Table 43 gives design information for trapezoidal-shaped outlets with varying peak flows and a maximum slope of 30 percent.

The riprap should grade gradually in diameter from the largest stones down to the smallest stones of 1-inch diameter. The diameter of the largest stones should be 1.5 times that of the average diameter stone. If the average stone diameter is 6.0 inches, the largest stone should be at least 9.0 inches in diameter. The thickness of the riprap layer should be at least 1.5 times the diameter of the maximum calculated stone diameter. If the largest stone is 9.0 inches in diameter, the riprap layer should be at least 13.5 inches thick. The riprap layer should be no less than 6 inches thick regardless of stone sizes.

Stone used for riprap should be a type that will not disintegrate on exposure to weather. This means that much of the shale and siltstone and some of the sandstone found on surface mines would not be suitable for use as riprap.

A more sophisticated method of determining the size of channel, shape of channel, and riprap size is described in Appendix F.

CHAPTER 16

SEDIMENTATION

It is usually not practical for the reclamationist to attempt total erosion control on a surface mine. The attempt would be extremely expensive and probably not successful. However, most of the sediment that does result from erosion must be held on the mine site. There are many methods and techniques for trapping sediment. Smaller amounts of sediment are often trapped by slowing water flow with dams made of bales of straw, stone dams, sandbag dams, ditches, and small excavated pits. These are excellent methods of holding sediment close to points where sediment is likely to originate. If the sediment is trapped before it joins other streams of moving sediment, there will be less sediment concentrating at the boundaries of the mine site. This can mean smaller sediment ponds, less sediment in the final sediment ponds, less sediment pond clean-out, and fewer citations from regulatory agencies.

The sediment being carried along by water runoff can be classified into three different types. The smallest particles of sediment are classified as wash load. These particles are clay and silt size. The smaller clay particles are colloidal in size and they often stay in suspension for long periods of time causing problems for the reclamationist. These colloidal particles stay in suspension because they have a negative electric charge and this charge causes the particles to repel each other. This repelling action causes the colloids to move constantly in all directions rather than falling to the bottom of the sediment pond. Certain chemicals can be added to water containing soil colloids that will cause the colloids to lose their negative charge and become attracted to each other. When the particles become attracted to each other, they cling together to form larger particles that will settle to the bottom of a sediment pond. The chemicals used to flocculate colloids are usually classified into three categories: (1) metal hydroxides such as calcium hydroxide; (2) metal salts such as aluminum sulfate; and (3) synthetic polymers that have a positive, negative, or neutral electric charge.

There is no accurate method of calculating the amount of flocculant to add to the sediment pond. At present most reclamationists collect a sample of the water containing the soil colloids and add flocculant until the soil colloids form particles large enough to settle. This method gives some indication of the amount and type of flocculant that will be needed. Field tests will still be needed to determine an optimum amount to be added to the sediment pond. If not flocculated, much of the colloidal material will pass through the pond or be deposited on the pond bottom if the pond dries.

The next larger sediment sizes are classified as suspended sediment load, and these particles can be fine sand, silt, or clay. The water must be moving to keep these particles in suspension. The largest particles are classified as bedload. These particles are not well suspended in the water. They move in the flowing water by sliding, bounding, or rolling along the bottom of the water channel.

The two factors that influence sedimentation are characteristics of the runoff water and characteristics of the soil particles being transported. Characteristics of the water that have an effect are the velocity of water flow and the turbulence of the water. As the velocity or the turbulence increases, there is an increase in the amount of sediment that can be carried by a given volume of water. The characteristics of the soil particles that affect sedimentation are size, shape, and density. Size is the characteristic that is most often considered and used in calculations, although shape and density do have an effect. A rough and irregularly shaped particle settles more slowly than a smooth and regularly shaped particle. A more dense particle settles more quickly than a less dense particle. Small particles such as clay, silt, and fine sand are more easily carried by water than the coarser sands and aggregated soil particles. The larger particles settle more quickly and are the first to be deposited when the velocity or turbulence of the water decreases.

SEDIMENT PONDS

The final structure in the erosion control chain is the sediment pond. The purposes of these ponds are to hold the runoff with suspended sediment long enough to let the sediment settle to the pond bottom, and to store the settled sediment.

There are two general types of sediment pond. One type, the dry pond, is built in a drainageway that carries water during and shortly after a rainfall. The other type, the wet pond, is formed by damming a permanent stream. The dry pond is probably preferable because it keeps permanent streams clean, is easier to dredge, and is less subject to damage. Figure 62 illustrates some of the features of a sediment pond. The dam is constructed of materials that allow a minimum amount of water seepage from the pond. Vegetation and riprap are used to prevent erosion of the dam. A pipe passes through the dam to allow water that has lost all or part of its sediment load to leave the pond by way of the

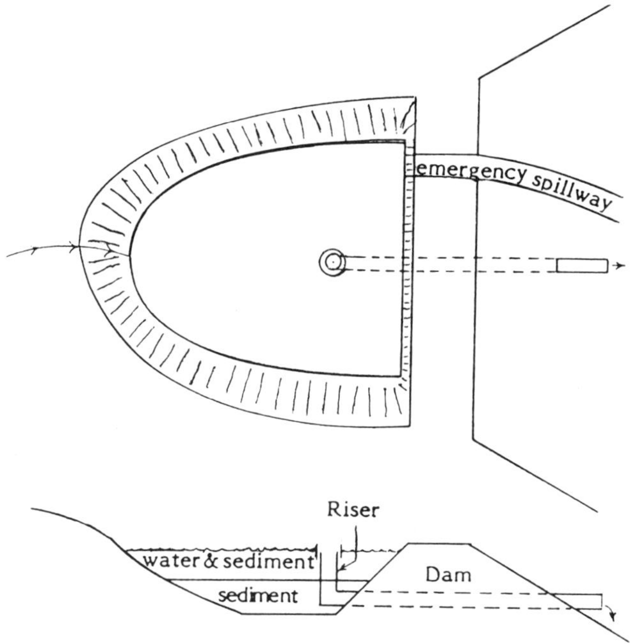

Figure 62. Top and side views of a sediment pond.

riser. An emergency spillway is provided to allow the passage of excess water that cannot be removed by the riser. This spillway prevents the erosion and loss of the dam during excessive rainfalls. The pond itself holds sediment laden water until part or all of the sediment is deposited on the pond bottom. The pond also acts as a storage reservoir for sediment.

It is difficult to give rules for locating sediment ponds on a surface mine because of the many factors that control location. In most cases the topography and mining operation will determine the pond location. After taking these circumstances into account, the ponds should be located to intercept the maximum amount of sediment-laden runoff. Accessibility for clean-out must also be taken into account, as many ponds must be dredged to keep them effective.

In many cases the same sediment-trapping operation can be accomplished with one large pond or two or more small ponds in series. An advantage of small ponds in series is that they are easier to build and maintain with smaller equipment. Another advantage is that the first pond in a series receives most of the water turbulence and sediment disturbance caused by inrushing storm water. Succeeding ponds receive less disturbance. This means that the effluent from the final pond should be relatively clean, even during peak runoff. A large pond is likely to require more elaborate construction, spillways, and outlets. The first of a series of small ponds, where sediment first enters, is likely to hold most of the sediment and it will be easier to clean with smaller equipment than a large pond. The succeeding small ponds will contain less and less sediment. Ponds can also be built in parallel. With such a system one pond can be cleaned while the other pond is in use.

SEDIMENT POND DESIGN

The design and construction regulations for sediment ponds can vary from state to state. Because of this circumstance, the following discussion will deal with the factors of sediment pond design but not with actual design or construction. Once the factors of design are understood, a pond can be designed and constructed to meet any practical regulatory requirement.

Sediment ponds can be designed to meet either of two primary objectives, or they can be designed to meet both objectives at the same time. The two objectives are:

1. To trap and hold a specified percentage of the sediment that enters the pond
2. To remove enough of the sediment to meet specified standards for sediment effluent from the pond

In order to meet either objective, appropriate dimensions for the pond must be determined. The volume of the pond is important for sediment storage purposes and the area of the water surface is important for sediment settling purposes. Water coming into a pond must travel far enough to the pond outlet to lose velocity and allow the sediment load to settle. Increasing the area of the pond decreases the velocity of the water flowing out of the pond. This allows smaller particles to settle and not leave the pond. The distance traveled by the water also provides the detention time needed to allow particles to settle the required distances. The following equation is used to determine the area of the pond:

$$A = \frac{Q}{V_S}$$

A = surface area of water in pond (cm^2)

Q = rate at which cleared water can be produced and, therefore, the rate at which water can be allowed to leave the pond (cm^3/s)

V_S = particle settling velocity (cm/s)

The value for settling velocity (V_S) is obtained by applying Stokes' law:

$$V_S = \frac{g}{18\mu}(S-1)D^2$$

Table 44. Soil Particle Size

Soil particle	Diameter range (cm)
Very coarse sand	0.2–0.1
Coarse sand	0.1–0.05
Medium sand	0.05–0.025
Fine sand	0.025–0.010
Very fine sand	0.010–0.005
Silt	0.005–0.0002
Clay	< 0.0002

V_S = settling velocity of particle (cm/s)

g = acceleration of particle due to gravity (981 cm/s^2)

D = diameter of particle (cm) (Table 44)

S = specific gravity of particle (2.65)

μ = kinematic viscosity of water (cm^2/s) (Table 45)

The use of this equation assumes that the particles are smooth and the water is not moving. Both of these assumptions probably will be false, but the values obtained are still useful. In order to apply this equation, the minimum size soil particle (D) that is to be removed, water viscosity (μ), and the specific gravity of the particle (S) must be known. The particle of smallest diameter that is practical to remove by settling is probably the smallest silt (0.0002 cm) (Table 44). These particles will settle about 17 cm (6.8 inches) in 24 hours in still water at 0°C (32°F). Most clay particles will take from 30 hours to several years to settle this distance. The specific gravity (S) for soil particles will be approximately 2.65, and this is the value often used in Stokes' equation. The final value needed in this equation is the kinematic viscosity of the water (μ) in the sediment pond. As water becomes colder, it becomes more viscous. This increase in viscosity means that a soil particle will settle more slowly. Under this circumstance it seems reasonable to use the viscosity that occurs at the average or even lowest winter temperature for the water in the sediment pond. Some kinematic viscosities for water at various temperatures are shown in Table 45.

All the values are now available for determining the settling rate (V_S) of soil particles in the sediment pond. Table 46 shows the settling velocities for various particle sizes at different water temperatures.

Example: Assume that the smallest particle to be removed from the runoff is the smallest silt fraction (0.0002 cm). The lowest winter temperature is 32°F and the specific gravity of the particles is 2.65.

Table 45. Kinematic Viscosity (μ) of Water at Various Fahrenheit and Celsius Temperatures

Temperature (°F)	Kinematic viscosity (μ) (cm^2/s)	Temperature (°C)	Kinematic viscosity (μ) (cm^2/s)
32	0.0179396	0	0.01792
40	0.0154591	5	0.01519
50	0.0130993	10	0.01308
60	0.0113063	15	0.01141
70	0.0098384	20	0.01007
89	0.0086399	25	0.00897
100	0.0076738	30	0.00804

Determine the length of time that the sediment will have to be detained by the settling pond in order to remove all sand and silt in the runoff.

$$V_s = \frac{g}{18\mu}(S-1)D^2$$

$$V_s = \frac{(981)}{(18)(0.0170)}(2.65-1)(0.0002)^2$$

$$V_s = 0.0002009 \text{ cm/s}$$

$$\text{24-h settling time x } V_s = \text{depth of settling in 24 h}$$
$$(86400 \text{ s}) \text{ x } (0.0002009 \text{ cm/s}) = 17.3577 \text{ cm (6.8 in.)}$$

This settling rate can now be used to determine the needed surface area (A) of the water in the pond. The equation for this surface area is:

$$A = \frac{Q}{V_s}$$

A = surface area of water in pond (m^2)

Q = rate of runoff for a predicted storm (m^3/s)

V_s = critical settling velocity (m/s)

The peak rate of runoff (Q) must be computed in order to use the equation, and the time period for computing the runoff is usually the expected runoff in a 10-year period for a 24-hour duration storm. The amount of runoff (R) must be computed first. The

Table 46. Settling Velocity (V_s) of Particles (cm/s) for Various Particle Diameters and Water Temperatures[1]

Temperature		Diameter (cm)							
(°F)	(°C)	0.2	0.1	0.05	0.025	0.01	0.005	0.0002	0.000100
32	0	200.50	50.12	12.53	3.13	0.50	0.12	0.000020	0.000050
40	4.4	232.68	58.17	14.54	3.64	0.58	0.14	0.00023	0.000057
50	10.0	274.56	68.64	17.16	4.29	0.69	0.17	0.00028	0.000070
60	15.6	318.04	79.51	19.88	4.97	0.79	0.20	0.00032	0.000080
70	21.1	365.56	91.39	22.85	5.71	0.91	0.23	0.00036	0.000090

[1]These settling velocities are for smooth particles in still water at a particle specific gravity of 2.65.

rational formula in a slightly different form than shown in the runoff section can be used.

$R = KPA$

R = amount of runoff (m^3)

K = runoff coefficient (Table 47)

P = rainfall (m)

A = runoff area (m^2)

Example: Assume that the mine is located in east Tennessee where the expected rainfall would be 5 inches in a 10-year period for 24-hour duration storm. The soil is a silt loam and the watershed consists of 10 acres of forest with an average slope of 20 percent, 6 acres of pasture with an average slope of 6 percent and 15 acres of minesoil with an average slope of 9 percent.

R for 10 acres of forest:

$K = 0.50$ (Table 47)

$P = 5$ in. x 0.0254 m/in. = 0.1270 m

$A = 10$ acres x 4047 m^2/A = 40470 m^2

$R = (0.50)\ (0.1270\ m)\ (40470\ m^2) = 2569.845\ m^3$ of runoff from forest

Table 47. Values for Coefficient K

	Soil texture and slope								
	Sandy loam			Clay loam and silt loam			Fine clay		
	(0–5)[1]	(5–10)	(10–30)	(0–5)	(5–10)	(10–30)	(0–5)	(5–10)	(10–30)
Forest	0.10	0.20	0.30	0.30	0.40	0.50	0.40	0.50	0.60
Pasture	0.10	0.20	0.30	0.30	0.40	0.50	0.40	0.50	0.60
Bare minesoil and cultivated land	0.30	0.40	0.50	0.50	0.60	0.70	0.60	0.70	0.80

[1]Numbers in parentheses indicate percent of slope.

Source: Frevert, R.K., et al. 1955. Soil and Water Conservation Engineering. New York: Wiley.

R for 6 acres of pasture:

$K = 0.40$ (Table 47)

$P = 5 \text{ in.} \times 0.254 \text{ m/in.} = 0.1270 \text{ m}$

$A = 6 \text{ acres} \times 4047 \text{ m}^2/A = 24282 \text{ m}^2$

$R = (0.40)(0.1270)(24282) = 1233.526 \text{ m}^3$ of runoff from pasture

R for 15 acres of minesoil:

$K = 0.60$ (Table 47)

$P = 5 \text{ in.} \times 0.0254 \text{ m/in.} = 0.1270 \text{ m}$

$A = 15 \text{ acres} \times 4047 \text{ m}^2/A = 60705 \text{ m}^2$

$R = (0.60)(0.1270)(60705) = 4625.721 \text{ m}^3$ of runoff from minesoil
$2569.845 \text{ m}^3 + 1233.526 \text{ m}^3 + 4625.721 \text{ m}^3 = 8429.092 \text{ m}^3$ of runoff expected in a 24-h period

To determine the rate of runoff for the entire period, the total runoff (8429.092 m^3) is divided by the number of seconds in 24 hours (86,400 s), to give $0.0975 \text{ m}^3/\text{s}$. However, this rate is probably much too slow, as most of the rainfall will take place within a short period of the 24 hours. The following equation is used to estimate the period of time when the greatest amount of runoff occurs.

$T = 6000\ (E_s - 0.20)$

T = duration intense rainfall (s)

E_s = total rainfall in inches for a 24-hour event based on a 10-year period

$T = 6000$ (5 in. - 0.20) = 28,800 s

The peak rate of runoff (Q) can now be estimated by dividing the total runoff (R) by the duration of intense rainfall (T).

$$Q = \frac{R}{T} = \frac{8429.092\ \text{m}^3}{28{,}800\ \text{s}} = 0.2927\ \text{m}^3/\text{s}$$

The surface area of water needed in the pond to remove all sediment 0.0002 cm in diameter or greater:

$$A = \frac{Q}{V_s} = \frac{0.2927\ \text{m}^3/\text{s}}{0.000002\ \text{m/s}} = 146{,}350\ \text{m}^2$$

$$146{,}350\ \text{m}^2 \div 4047\ \text{m}^2/A = 36.16\ \text{acres}$$

The minimum depth of the water-holding portion of this pond would be:

$$D = \frac{\text{peak runoff } (R)}{\text{surface area of water in the pond } (A)}$$

$$D = \frac{8429.09\ \text{m}^3}{146{,}350\ \text{m}^2} = 0.0576\ \text{m}$$

The area of this pond is too large for most surface mine situations. Also the shallow depth would allow scouring and resuspension of the sediment that has been deposited on the bottom of the pond. If the smallest diameter particle to be trapped is very fine sand (diameter = 0.005 cm), the size of the pond can be reduced drastically.

V_s = 0.125593 cm/s for particles of 0.005-cm diameter

$$A = \frac{Q}{V_s} = \frac{0.2927\ \text{m}^3/\text{s}}{0.00125593\ \text{m/s}} = 233.05439\ \text{m}^2$$

$$233.05439\ \text{m}^2 \div 4047\ \text{m}^2/A = 0.05758\ \text{acres}$$

$$D = \frac{R}{A} = \frac{8429.09\ \text{m}^3}{233.05\ \text{m}^2} = 36.17\ \text{m}$$

All calculations up to this point have been based on an ideal settling velocity of smooth spherical particles, in still water, with controlled movement of incoming water to the outlet, no dead areas, and several other factors needed for ideal settling. In order to compensate for the disrupting factors that will actually occur, the final surface area should be increased by a factor of 1.2. This means that the surface area of the preceding pond would be:

$$233.05\ \text{m}^2 \times 1.2 = 279.66\ \text{m}^2$$

$$279.66\ \text{m}^2 \div 4047\ \text{m}^2/\text{acre} = 0.06901\ \text{acres}$$

The depth can now be decreased and obtain the same runoff storage volume:

$$8429.09 \text{ m}^3 \text{ of runoff } \div 279.66 \text{ m}^2 \text{ of area} = 30.14 \text{ m (98.89 ft) of depth}$$

The pond is now small in surface area (55 ft x 55 ft), but excessively deep (99 ft). Under these circumstances it would be practical to increase the surface area by utilizing any space that is available to increase the length and width of the pond. Assume that there is sufficient space to build a pond that is 100 ft (30.48 m) wide and 200 ft (60.96 m) long. This would give a pond with a surface area of 1858.06 m^2. The new pond area would be:

$$8429.09 \text{ m}^3 \text{ of runoff } \div 1858.06 \text{ m}^2 \text{ of area} = 4.54 \text{ m (14.88 ft) of depth}$$

This new depth would be more practical and the increased surface area would decrease water turbulence, increase time for sediment settling, and remove particles smaller than very fine sand.

The pond depth under discussion is the depth needed to contain the calculated body of water. This depth should be increased to compensate for dam settling, wave action, errors, and unforeseen circumstances. This increase in depth is called freeboard and is the vertical distance from the top of the dam to the maximum expected water elevation. A pond with a length of 500 feet or less should have a 1-foot freeboard. A pond with a length between 500 and 1000 feet should have a 1.5-foot freeboard and ponds over 1000 feet in length should have a 2-foot freeboard.

SEDIMENT STORAGE

Up to this point, pond storage for sediment-laden water has been considered. Now the additional storage for sediment alone will be considered. The simplest method of determining the volume of sediment storage needed in the sediment pond is to assume that a standard volume of sediment will be contributed by each disturbed acre. Some regulatory authorities solve the problem by requiring that ponds be designed to store a given amount of sediment for each acre of watershed. The authority then requires that the sediment be removed when the total pond volume is reduced by sediment to a specified minimum.

It would be relatively simple if the gross erosion as predicted by the universal soil loss equation could be used to determine the volume of sediment to be stored. However, only a part of this gross erosion will reach the pond. Some will be deposited before it reaches the pond. This problem is solved by applying a ratio of sediment yield to gross erosion. This ratio is termed the sediment delivery ratio. The other problem in determining the amount of sediment to be stored is accounting for the

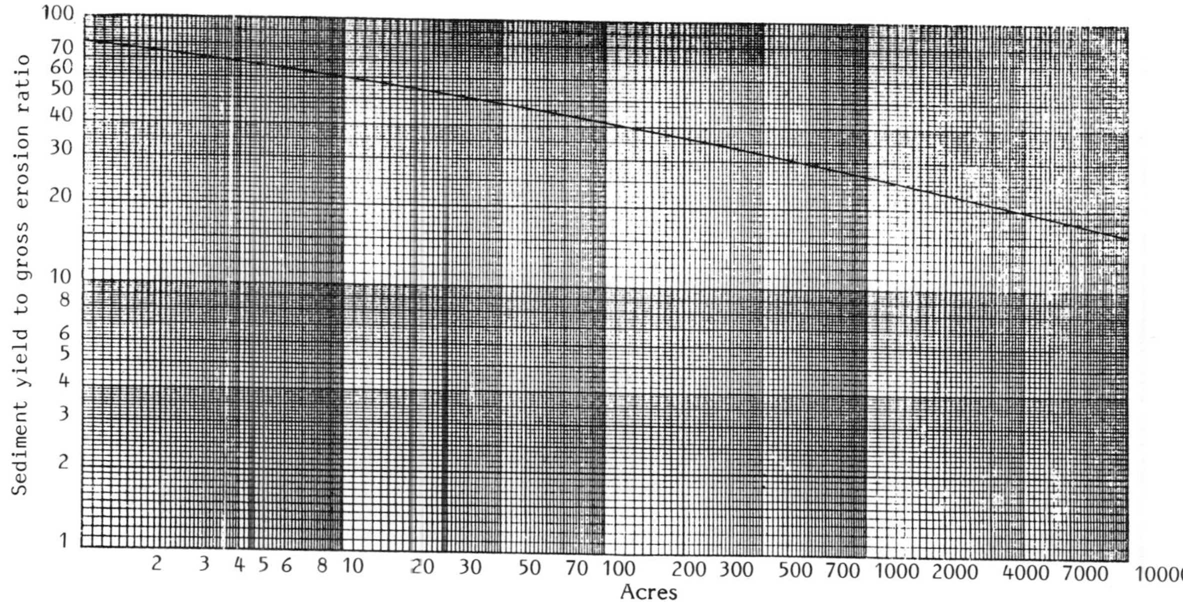

Figure 63. Effect of watershed size on sediment yield ratio. *Source:* U.S. Department of Agriculture, Soil Conservation Service. 1971. Sediment Sources, Yields and Delivery Ratios. *National Engineering Handbook.* Section 3, Sedimentation; Chapter 6. Washington, D.C.: U.S. Government Printing Office.

Table 48. U.S. Soil Conservation Service Values for Specific Weights of Sediments in Ponds

Textures	Submerged sediment[1] (lb/ft^3)	Aerated sediment[1] (lb/ft^3)
Clay	40–60	60–80
Silt	55–75	75–85
Clay-silt mixture	40–65	65–85
Sand-silt mixture	75–85	95–110
Clay-silt-sand mixture	50–80	80–100
Sand	85–100	85–100

[1]Finer textures develop higher specific weights than coarser textures when allowed to dry.

sediment that passes through the pond. The trap efficiency of the pond is used to account for this sediment loss. Trap efficiency is controlled by detention times and velocity of outflow from the pond. As detention time increases and as outflow velocity decreases, the trap efficiency increases.

The U.S.D.A. Soil Conservation Service has developed a rule of thumb stating that sediment ponds for well-vegetated areas should be designed to store a minimum of 0.01 acre-foot (435 cubic feet) per acre of watershed per year. An unvegetated area such as a newly topsoiled surface mine would need a pond designed to store 0.06 acre-foot (2614 cubic feet) per acre of watershed per year. If the bare watershed is vegetated or mulched immediately, the storage capacity can be reduced to 0.045 acre-foot (1960 cubic feet) per acre of watershed per year. This last situation should apply to most surface mine reclamation operations.

A somewhat more sophisticated and complicated system for estimating sediment yield has been developed. In this system the ratio of sediment yield to gross erosion is correlated with watershed area. In order to use this system, the gross erosion and watershed area must be known. For newly regraded surface mines the gross erosion can be approximated by applying the universal soil loss equation. Approximate textures can be determined by feel or obtained from county soil survey manuals and watershed area can be obtained from maps. To use this system, the watershed area is applied to Figure 63. The sediment delivery to gross erosion ratio is found, and the ratio value is applied in the following equation:

$$Y = ER$$

Y = sediment yield (tons)

E = gross erosion (tons per watershed per year)

R = ratio of sediment yield to gross erosion

Example: Assume that a 32-acre watershed is emptying into a sediment pond and the universal soil loss equation predicts that 160 tons of soil will be eroded each year from the 32 acres. Figure 63 shows that a 32-acre watershed would have an approximate ratio of 0.50.

$$Y = ER$$

Y = 160 tons x 0.50

= 80.0 tons of sediment

Assume that the sediment is practically all sand and silt. The weight of this sediment will be approximately 100 lb/ft^3 (Table 48). Therefore, a ton of this sediment would have a volume of 20 ft^3 (2000 lb of sediment ÷ 100 lb/ft^3 of sediment = 20 ft^3), and 80.0 tons of sediment would require a storage volume of 1600 ft^3 (20 ft^3/ton x 80.0 tons = 1600 ft^3).

SEDIMENT POND TRAP EFFICIENCY

Usually the sediment pond does not trap all the sediment that enters. Some sediment will probably be carried out of the pond by outflow water. The ability of a pond to trap and hold sediment is called trap efficiency. Trap efficiency is the percent of entering sediment that remains in the pond as deposited sediment.

The two most important factors controlling trap efficiency are the sediment particle size and the length of time that the sediment-laden water stays in the pond. Larger particles settle from the water more quickly than smaller particles. Therefore a coarse-textured sediment will be deposited on the pond bottom before a fine-textured sediment. A pond that holds the sediment-laden water for a long period allows more of the finer sediment to deposit than a pond that holds the incoming water for only a short time.

There is no accurate method for estimating the trap efficiency of a sediment pond. However, a useful method has been developed based on the ability of a pond to retain water long enough to deposit significant amounts of the sediment load. Figure 64 is a set of curves developed by G.M. Brune to show the relationship between inflow, pond capacity, and

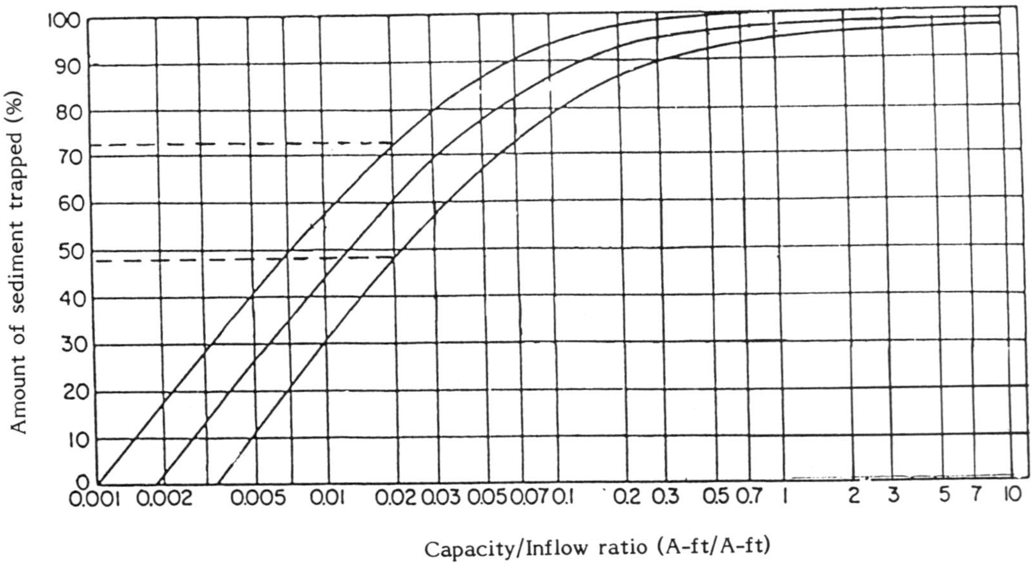

Figure 64. The percent of sediment that would be trapped by a pond of known capacity (acre-feet) and known annual water inflow (acre-feet). *Source:* Brune, G.M. 1953. Trap Efficiency of Reservoirs. *Trans. Am. Geophys. Union* 34:407–418.

trap efficiency. To use these curves, the capacity of the pond and the amount of water flowing into the pond each year must be known.

> **Example:** A pond has a capacity of 3.44 acre-feet and the annual water inflow is 172 acre-feet per year. The capacity/inflow ratio will be 0.02. Figure 64 indicates that the average amount of sediment trapped by such a pond will be about 61 percent of the sediment that enters the pond. The maximum and minimum amounts of sediment trapped will be approximately 72 percent and 48 percent respectively.

Remember that this is only an estimate. The actual amount of sediment trapped will be controlled by many factors that have not been considered. The sediment particle size has a great effect on amount of sediment trapped. Coarse particle size such as sand or aggregated clays will fall to the bottom of the pond quickly. Silt and clay will remain in suspension for longer periods of time and will be more likely to pass through the pond. The amount of time that sediment-laden water stays in the pond will affect the amount of sediment trapped. This time factor is associated with the shape, size, and outlet location of the pond. The size of the outlet can also determine the length of time that water remains in a pond.

REFERENCES

Beasley, R.P. 1972. *Erosion and Sediment Pollution Control.* Ames, IA: Iowa State University Press.

Brune, G.M. 1953. Trap Efficiency of Reservoirs. *Trans. Am. Geophys. Union.* 34(3):407-418.

Frevert, R.K., G.O. Schwab, T.W. Edminster, and K.K. Barnes. 1955. *Soil and Water Conservation Engineering.* New York: Wiley.

Hershfield, D.M. 1961. *Rainfall Frequency Atlas of the United States,* U.S. Weather Bureau Technical Paper 40. Washington, D.C.

Kirpich, P.Z. 1940. Time of Concentration of Small Agricultural Watersheds, *Civil Engineering* 10(6).

Phillips, R.L. 1972. Design, Layout, Construction and Maintenance of Terraces. *American Society of Agricultural Engineers Yearbook,* Recommendation R 268. St. Joseph, MI.

U.S. Department of Agriculture, Soil Conservation Service. 1971. Sediment Sources, Yields and Delivery Ratios. *National Engineering Handbook.* Section 3, Sedimentation; Chapter 6. Washington, D.C.

U.S. Department of Agriculture, Soil Conservation Service. 1975. *Urban Hydrology for Small Watersheds,* Technical Release No. 55. Washington, D.C.: U.S. Government Printing Office.

Wischmeir, W.H. and D.D. Smith. 1978. *Predicting Rainfall Erosion Losses: A Guide to Conservation Planning,* Agriculture Handbook No. 537, U.S. Dept. of Agriculture, Science and Education Administration. Washington, D.C.: U.S. Government Printing Office.

SUGGESTED READING: PART III

Beasley, R.P., J.M. Gregory, and T.R. McCarty. 1984. *Erosion and Sediment Pollution Control.* 2nd ed. Ames, IA: The Iowa State University Press.

Chow, V.T. 1964. *Handbook of Applied Hydrology.* New York: McGraw-Hill.

Hudson, N. 1981. *Soil Conservation.* 2nd ed. Ithaca, NY: Cornell University Press.

U.S. Department of Agriculture, Soil Conservation Service. 1973. *S.C.S. National Engineering Handbook,* Sections 3 and 4. Washington, D.C.: U.S. Government Printing Office.

U.S. Department of Agriculture, Soil Conservation Service. 1979. *S.C.S. Engineering Field Manual for Conservation Practices,* Chapters 2-4, 6-11, 16. Washington, D.C.: U.S. Government Printing Office.

U.S. Department of Agriculture, Soil Conservation Service. 1975. *Urban Hydrology for Small Watersheds,* Technical Release No. 55. Washington, D.C.: U.S. Government Printing Office.

U.S. Department of the Interior, Office of Surface Mining. 1982. *Surface Mining Water Diversion Design Manual.* Washington, D.C.: U.S. Government Printing office.

U.S. Environmental Protection Agency. 1976. *Erosion and Sediment Control—Surface Mining in the Eastern United States.* Report No. EPA 625/3-7-006. U.S. Environmental Protection Agency. Washington, D.C.: U.S. Government Printing Office.

Wischmeier, W.H., and D.D. Smith. 1978. *Predicting Rainfall Erosion Losses: A Guide to Conservation Planning,* Agriculture Handbook No. 537, U.S. Dept. of Agriculture, Science and Education Administration. Washington, D.C.: U.S. Government Printing Office.

APPENDIXES

APPENDIX A

SEED, SOWING, AND PLANT INFORMATION

Seed Information

Plant			Seeding rate per acre	
No.	**Common name**	**Latin name**	**Drill (lb)**	**Broadcast (lb)**
1	Alfalfa	*Medicago sativa*	15	22
2	Bahiagrass	*Paspalum notatum* Flugge	12	20
3	Barley	*Hordeum vlugare* L.	80	100
4	Common Bermuda grass	*Cynodon dactylon* (L). Pers.		
		Hulled	8	10
		Unhulled	15	15
5	Birdsfoot trefoil	*Lotus corniculatus* L.	5	10
6	Black medic	*Medicago lupulina* L.	10	15
7	Annual bluegrass	*Poa annua* L.	20	25
8	Canada bluegrass	*Poa compressa* L.	20	25
9	Kentucky bluegrass	*Poa pratensis* L.	20	30
10	Big bluestem	*Andropogon furcatus* Muhl.	15	15
11	Little bluestem	*Andropogon scoparius* Michx.	15	15
12	Bromegrass, smooth	*Bromus inermis* Michx. Leyss.	15	20
13	Common buckwheat	*Fagopyrum saggitatum* Grlib	36–60	--
14	Buffalo grass	*Buchloe dactyloides*	5	8
15	Carpet grass	*Axonopus compressus* (swertz. Beauv.)	12	12

continued

Seed Information *(Continued)*

Plant			Seeding rate per acre	
No.	**Common name**	**Latin name**	**Drill (lb)**	**Broadcast (lb)**
16	Chufa	*Cyperus esculentus* L.	15–40 (nuts)	—
17	Clover alsike	*Trifolium hybridum* L.	6	8
18	Clover, crimson	*Trifolium incarnatum* L.		
		Hulled	15	30
		Unhulled	50	60
19	Clover alyce	*Alysicarpus vaginalis* DC.	10	20
20	Clover, bur	*Medicago hispide* Gaertn.	15	25
21	Clover, hop	*Trifolium agrarium* L.	4	6
22	Clover, ladino or white	*Trifolium repens* L.	3	7
23	Clover, red	*Trifolium pratense* L.	6	15
24	Clover, rose	*Trifolium hirtum*	10	20
25	Clover (white sweet)	*Melilotus alba* Med.		
		Unhulled	15	25
		Hulled	10	20
26	Clover (yellow sweet)	*Melilotus officinalis* Lam.		
		Unhulled	15	25
		Hulled	10	20
27	Clover, arrowleaf	*Trifolium vesiculosum* Savi	5	8
28	Clover, subterranean	*Trifolium subterraneum* L.	8	20
29	Crotalaria, showy	*Crotalaria spectabilis* Roth.	15	25
30	Crownvetch	*Coronilla varia* L.	5	15
31	Flatpea	*Lathyrus sylvestris*	15	30
32	Dallisgrass	*Paspalum dilatatum* Poir.	8	20
33	Deertongue	*Panicum clandestinum*	8	15
34	Fescue, tall	*Festuca arundinacea*	10	15
35	Grama, sideoats	*Bouteloua curtipendula*	10	20
36	Indiangrass	*Sorghastrum nutans*	5	15
37	Johnsongrass	*Sorghum halepense* (L.) Pers.	10	20
38	Kleingrass	*Panicum coloratum* L.	—	—
39	Lespedeza, sericea	*Lespedeza cuneata* (Dumde Cours) G.		
		Unscarified or unhulled	10	15
		Scarified or hulled	10	15
40	Lespedeza, Kobe	*Lespedeza striata* Hook & Arm	5	15

continued

Seed Information *(Continued)*

Plant			Seeding rate per acre	
No.	**Common name**	**Latin name**	**Drill (lb)**	**Broadcast (lb)**
41	Lespedeza, common	*Lespedeza striata* Hook & Arm	5	10
42	Lovegrass, weeping	*Eragrostis curvula* schrad. Nees	1/4	1/2
43	Lupine, blue	*Lupinus angustifolius* L.	70	100
44	Lupine, white	*Lupinus albus* L.	90	150
45	Lupine, yellow	*Lupinus luteus* L.	45	55
46	Millet, browntop	*Panicum ramosum* L.	10	25
47	Millet, foxtail	*Setaria italica* (L.) Beauv.	20	30
48	Millet, Japanese (barnyard)	*Echinochloa crusgalli frumentacea* W.F. Wight	20	30
49	Millet, pearl	*Pennisetum glaucum* L.	16	25
50	Millet, proso	*Panicum miliaceum* L.	15	90
51	Oats, common	*Avena sativa* L.	70	90
52	Oatgrass, tall	*Arrhenatherum elatius*	30	40
53	Orchardgrass	*Dactylis glomerata* L.	20	20
54	Pangolagrass	*Digitaria decumbens*	Establish vegetatively	
55	Redtop	*Agrostis alba* L.	8	10
56	Rhodesgrass	*Chloris gayana* Kunth	8	15
57	Reed canarygrass	*Phalaris arundinacea* L.	8	10
58	Rescuegrass	*Bromus catharticus* Vahl.	25	25
59	Sudangrass	*Sorghum volgare sudanense* (Piper) Hitchc.	20	25
60	Rye	*Secale cereale* L.	90	100
61	Ryegrass, annual	*Lolium multiflorum* Lam.	25	30
62	Ryegrass, perennial	*Lolium perenne* L.	25	30
63	Sesbania	*Sesbania macrocarpa* Muhl.	20	25
64	Sorghum	*Sorghum bicolor* L. Moench	15	20
65	Sorghum-Sudan		20	30
66	Switchgrass	*Panicum virgatum* L.	5	10
67	Timothy	*Phleum pratense* L.	8	8
68	Wheat, common	*Triticum vulgare* Vill.	60	70
69	Wheatgrass, western	*Agropyron smithii*	10	15
70	Vetch, common	*Vicia sativa* L.	40	60
71	Vetch, hairy	*Vicia villosa* Roth	40	50

Sowing Information

Plant no.	Seeds per pound (thousands)	Minimum germination needed (percent)	Seed weight per bushel (lb)	Time for germination (days)	Growth season[1]	Growth habit[2]	Depth of planting (in.)
1	220	80	60	7	C	P	0–1/4
2	155	85	60	21	W	P	1/4–1/2
3	13	85	48	7	C	A	1–2
4	1800	85	40	21	W	P	1/4–1/2
	1800	85	40	21	W	P	1/4–1/2
5	375	80	60	7	CW	P	0–1/4
6	300	80	60	7	C	A	0–1/4
7	1200	75		21	C	A	1/4–1/2
8	2500	75	14	28	C	P	1/4–1/2
9	2200	80	14	28	C	P	0–1/4
10	150	60		28	W	P	1/2–3/4
11	260	60		28	W	P	1/2–3/4
12	137		14	14	C	P	0–1/2
13	20		48	6	W	A	1 1/2–2
14	56			28	W	P	1–1 1/2
15	1350	85	18–36	21	W	P	1/4–1/2
16			44		W	P	2–3
17	680	85	60	7	C	P	1/4–1/2
18	150	85	60	21	W	WA	0–1/2
	150	85	60	21	W	WA	0–1/2
19	275	85	60	21	W	A	1/4–1/2
20	209	85	50	14	C	WA	1/2–3/4

21	830	85	60		C	WA	0–1/4
22	860	85	60	7	C	P	0–1/4
23	250	80	60	7	C	P	1/4–1/2
24	140	85	60				
25	250	80	60	7	C	B;A	1/4–1/2
	250	80	60	7	C	B;A	1/4–1/2
26	250	80	60	7	C	B	1/4–1/2
	250	80	60	7	C	B	1/4–1/2
27	400	85			C	A	1/8–1/4
28	65	85	60	14	C	WA	0–1/2
29	30	75	60	10	W	A	1–2
30	140	65	80		C	P	1/4–1/2
31					W	P	
32	340	60	15	21	W	P	1/4–1/2
33					W	P	
34	227	80			C	P	1/4–1/2
35	190	65		28	W	P	1/4–1/2
36	175	65			W	P	0–1/2
37	130	60	28	35	W	P	1/2–1
38					W	P	
39	372		35		W	P	0–1/4
	335		60	28	W	P	0–1/4
40	185	85	25	14	W	A	0–1/4
41	343	85	25	14	W	A	0–1/4
42	1500	80		14	W	P	0–1/4
43	3	80	60	7	C	A	1–2
44	3	80	60	7	C	A	1–2
45	4	80	60	21	C	A	1–2
46	140	80		14	W	A	0–1/2
47	220	85	50	10	W	A	1/2–3/4
48	155		35	10	W	A	0–1/2

continued

Sowing Information *(Continued)*

Plant no.	Seeds per pound (thousands)	Minimum germination needed (percent)	Seed weight per bushel (lb)	Time for germination (days)	Growth season[1]	Growth habit[2]	Depth of planting (in.)
49	85	80		7	W	A	3/4–1
50	80	80	56	7	W	A	3/4–1
51	14	85	32	10	C	A	1–2
52	150	50	11–14	14	C	P	0–1/2
53	590	80	14	18	C	P	0–1/4
54					W	P	
55	5100	85	14	10	C	P	0–1/4
56	2140	60	8–12	14	W	P	0–1/4
57	550	80	45	21	C	P	0–1/4
58	70	80	10	35	W	B,WA	3/4–1
59	55	80	40	10	W	A	3/4–1
60	18	85	56	7	C	A	1–2
61	227	85	24	14	C	A	0–1/4
62	330	85	60	7	W	P	0–1/4
63	40	85	55	10	W	A	1/2–1
64	20	80	45	10	W	A	1–2
65	35	80	45	10	W	A	1–2

66	389	75		28	W	P	0-1/4
67	1230	85	60	10	C	P	0-1/4
68	12	85	60	7	C	A	1-2
69	110	80			C	P	
70	7	85	60	10	C	A	1 1/2-2 1/2
71	20	85			C	A	1-2

[1]C = cool; W = warm.

[2]P = perennial; A = annual; WA = reseeding annual; B = biennial.

Plant Information

Plant no.	Regions where grown (see Figure A-1)	When planted	Soil[1]
1	All except 2b	Spring or late fall	Alkaline, pH 6.5–8.5, well-drained
2	2a, 2b	Spring	Moist to droughty, pH 4.5–7.5
3	All	Spring, fall or winter	Dry, medium texture, pH 5.5–8.0, moderate salt content
4	2a, 2b	Spring	pH 5.0–7.0, moderate salt content, coarse to fine texture
	2a, 2b	Spring	pH 5.0–7.0, moderate salt content, coarse to fine texture
5	1a, 1b	Spring	Acid, wet, pH 5.0–7.5
6	All	Fall	Alkaline
7	1a	Fall or spring	pH 5.5–7.0
8	1a	Fall or spring	pH 4.5–7.0, droughty, infertile
9	1a, 1b	Early fall/spring	pH 5.5–7.0, well-drained
10	1a, 1b, 3a, 3b	Spring	pH 5.0–7.5
11	1a, 1b, 3a, 3b	Spring	pH 6.0–8.0
12	1a, 1b, 3a	Spring or fall	Coarse to fine texture, low salt content, pH 5.5–8.0
13		12 weeks before first fall frost	pH 5.0–7.5, coarse texture, acid, well-drained
14	3a, 3b	Spring	pH 6.5–8.0, medium to fine texture
15	2a, 2b	Winter or spring	pH 4.7–7.1, coarse texture
16	2a, 2b	Spring and summer	
17	1a, 1b, 3a	Spring or fall	Low salt content, pH 5.2–7.0, wet
18	1b, 2a, 2b	Late summer, early fall	Well-drained, pH 5.5–7.0, avoid alkaline soils, coarse to fine texture
	1b, 2a, 2b	Late summer, early fall	Well-drained, pH 5.5–7.0, avoid alkaline soils, coarse to fine texture
19	2b	Spring	Well-drained
20	2a, 2b, 3b	Fall	Alkaline
21	1a		
22	All	Spring or fall	Low salt content, ph 6.0–7.5, moist soils

continued

Plant Information *(Continued)*

Plant no.	Regions where grown (see Figure A-1)	When planted	Soil[1]
23	All except 2b	Spring, late fall, or summer	Low salt content, pH 5.5-7.0, well-drained
24			
25	All except 2b	Early spring	Alkaline, pH 6.0-8.0, moderate salt content
	All except 2b	Early spring	Alkaline, pH 6.0-8.0, moderate salt content
26	All except 2b	Early spring	Alkaline, pH 6.0-8.0
	All except 2b	Early spring	Alkaline, pH 6.0-8.0
27	2a	Late summer or early fall	Well-drained
28	2a, 2b	Fall	Well-drained
29	2a, 2b, 1b	Spring	Coarse texture, well-drained, acid
30	1a, 1b, 3a	Fall	Well-drained, pH 5.5-7.5
31	1a, 1b, 2a	Spring	Droughty, pH 7.0-4.5
32	2a, 2b	Late fall	Moist, fine texture
33	1a, 1b, 2a, 3b	Fall, winter, spring	pH down to 4.0
34	1a, 1b	Late summer or early fall	pH 5.4-8.0, moderate salt content, moist
35	1b, 3b	Spring	pH 4.5-7.0
36	1a, 1b, 2b	Spring	pH 4.5-7.0
37	2a, 2b	Spring or summer	Alkaline, medium texture
38	Texas		
39	1b, 2a	Early spring	Well-drained, avoid alkaline soils, pH 5.0-7.0
	1b, 2a	Early spring	Well-drained, avoid alkaline soils, pH 5.0-7.0
40	1b, 2a, 2b	Spring	Avoid alkaline soils, well-drained, pH 5.0-6.0
41	1b, 2a, 2b	Spring	Avoid alkaline soils, well-drained, pH 5.0-6.0
42	1b, 2a, 3b		pH 4.5-8.0, infertile soils
43	2a, 2b	Fall	Coarse texture, well-drained
44	2b	Fall	Coarse texture, well-drained
45	2a, 2b	Fall	Coarse texture, well-drained
46	2a	May-July	Well-drained
47	3a, 3b	Spring	Well-drained, pH 4.5-7.0

continued

Plant Information *(Continued)*

Plant no.	Regions where grown (see Figure A-1)	When planted	Soil[1]
48	All	Spring, early summer	Wet soils, pH down to 4.5
49	2a, 2b	Late spring	Well-drained, avoid alkaline soils
50	3a, 3b	Late spring	Well-drained, coarse to fine texture
51	All	Spring or fall	Well-drained, pH 5.0-8.0, medium to fine texture
52	1a, 1b		pH 5.0-7.5
53	1a, 1b	Spring or fall	Well-drained, medium to fine texture, pH 5.0-7.5
54	2a, 2b	Spring	Well-drained, droughty
55	1a, 1b	Early spring or late summer	Acid, pH 4.0-7.6, wet
56	2b, 3b	Spring	Salt tolerant
57	1a, 1b	Early spring	Moist to wet, pH 5.0-7.5
58	2a, 2b	Fall	Moderate salt content
59	All	Spring or summer	Dry, alkaline, coarse to fine texture, pH 5.5-7.5
60	All	Spring or fall	Acid pH 5.0-8.0, coarse to medium texture, well-drained
61	1a, 1b, 2a	Fall or spring	pH 5.5-8.0, medium texture
62	1a, 1b	Fall/early spring	pH 5.5-8.0, medium texture
63	2a, 2b, 3b	Spring	
64	2a, 1b, 1a, 3b, 3a	Spring or summer	Dry, moderate salt content, pH 5.5-8.0, coarse texture
65	1a, 1b, 2a, 3a, 3b	Spring or summer	pH 5.6-6.5, well-drained
66	1a, 1b, 3a, 3b	Spring	pH 4.5-7.5, low fertility
67	1a, 1b	Spring or fall	Medium to fine texture, pH 4.5-8.0, well-drained

continued

Plant Information *(Continued)*

Plant no.	Regions where grown (see Figure A-1)	When planted	Soil[1]
68	All	Fall	Medium to fine texture, pH 5.5-8.0
69	3a, 3b		pH 4.5-7.0, drought resistant
70	2a, 1b	Spring	Coarse to fine texture, pH 5.5-6.5, well-drained
71	1b, 2a	Fall	pH 5.0-7.5, coarse to fine texture

[1]Soil characteristics to be considered in growing each species. The soil characteristics described are those best for growth or those that the species will tolerate.

Figure A-1. Regions where species described in Appendix A—Plant Information are likely to make their best growth.

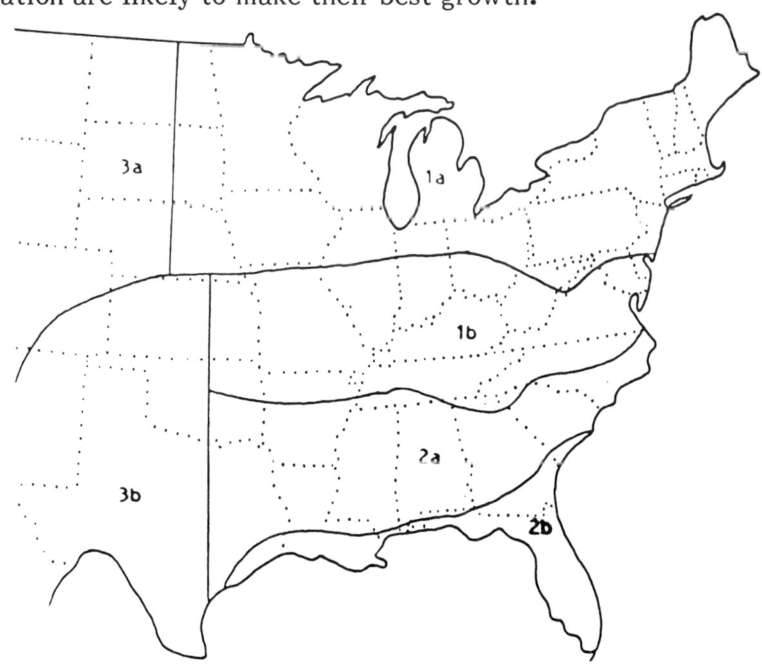

APPENDIX B

TREE SEED SIZE, TREATMENT, PLANTING, AND STORAGE

Tree		No. of seeds per pound	Maximum planting depth (in.)	Temperature for short-term storage (°F)	Treatment to break dormancy[1]
Latin name	Common name				
Acer rubrum	Red maple	12,700-38,200	0.50	35-41	2
Abies balsamea	Balsam fir	30,000-94,500	0.25	35	2
Ailanthus altissima	Tree-of-heaven	13,300-17,400	0.50	41	2
Alnus glutinosa	European black alder	260,000-640,000	0.25	35	8
Betula lenta	Sweet birch	443,000-933,000	0.05	35	2
Carya illinoensis	Pecan	55-160	1.00	41	2
Celtis occidentalis	Hackberry	3,500-5,400	0.50	41	2
Cercis canadensis	Redbud	14,000-25,000	0.15	35	6,7
Chamaecyparis thyoides	Atlantic white-cedar	420,000-500,000	0.15	35	1
Cornus florida	Flowering dogwood	3,300-6,200	0.25	38-41	4
Cupressus arizonica	Arizona cypress	30,000-90,000	0.15	34-41	9
Diospyros virginiana	Persimmon	665-1,764	2.00	37	2
Eleagnus umbellata	Autumn olive	20,000-54,000	0.50	34	2
Fagus grandifolia	American beech	1,300-2,300	0.50	35	2
Fraxinus pennsylvania	Green ash	11,000-35,000	0.50	41	2
Gleditsia triacanthos	Honeylocust	1,750-4,050	0.50	32-45	6
Juglans nigra	Black walnut	32-96	1.50	37	4
Juniperus virginiana	Eastern redcedar	37,000-55,000	0.25	30	4
Larix laricina	Tamarack	210,000-420,000	0.15	20	1
Lespedeza bicolor	Bicolor lespedeza	64,000-85,000	0.25	50	6
Liquidambar styraciflua	Sweetgum	65,000-98,000	0.15	35-40	2
Liriodendron tulipifera	Yellow-poplar	5,000-30,000	0.25	36-40	2
Lonicera tatarica	Tatarian honeysuckle	116,000-198,000	0.15	34-38	2

Magnolia grandiflora	Southern magnolia	5,800-6,800	0.25	32-41	4
Malus pumila	Apple	7,000-27,000	0.75	36-50	2
Morus rubra	Red mulberry	200,000-500,000	0.15	10-0	2
Nyssa sylvatica var. sylvatica	Blackgum	2,500-4,000	0.25	33	2
Paulonia tomentosa	Royal paulonia	2,800,000	0.00	35-40	1
Picea rubens	Red spruce	100,000-289,000	0.25	33-38	2
Picea mariana	Black spruce	335,000-510,000	0.15	33-38	2
Pinus palustris	Longleaf pine	3,000-7,000	0.00	0-41	7
Pinus rigida	Pitch pine	42,000-82,000	0.15	0-41	7
Pinus strobus	Eastern white pine	17,500-53,000	0.15	0-41	2
Pinus taeda	Loblolly pine	12,300-26,400	0.00	0-41	2
Pinus virginiana	Virginia pine	45,700-91,000	0.25	0-41	7
Platanus occidentalis	American sycamore	100,000-250,000	0.25	20-38	1
Populus deltoides var. deltoides	Eastern cottonwood	200,000-600,000	0.00	41	1
Populus grandidentata	Bigtooth aspen	3,000,000+	0.00	41	1
Prunus serotina	Black cherry	2,800-13,800	1.00	33-41	4
Quercus acutissima	Sawtooth oak	94-110	0.50	32-36	1
Quercus alba	White oak	70-210	0.50	32-36	1
Quercus palustris	Pin oak	320-540	0.50	32-36	2
Quercus rubra	Northern red oak	74-250	0.50	32-36	2
Robinia pseudoacacia	Black locust	24,000	0.25	32-40	6
Robinia hispida	Rose-Acacia locust	27,000	0.25	32-40	6
Robinia fertilis	Bristly locust	32,000	0.25	32-40	6

continued

(Continued)

Tree		No. of seeds per pound	Maximum planting depth (in.)	Temperature for short-term storage (°F)	Treatment to break dormancy[1]
Latin name	Common name				
Salix nigra	Black willow	2,500,000	0.00	—	1
Tilia americana	American basswood	5,000	0.25	33-40	3
Tsuga canadensis	Eastern hemlock	124,000-360,000	0.00	41	2
Ulmus americana	American elm	50,000-100,000	0.25	25-40	2

[1] 1 = No treatment
2 = Stratify (seed stored under moist conditions at a low temperature) for 2 to 3 months at 33 to 41°F.
3 = Soak in concentrated nitric acid for 1/2 to 2 hours, then stratify for 3 to 5 months at 33 to 41°F.
4 = Stratify for 4 months at 33 to 41°F.
5 = Stratify for 2 to 4 months at 68 to 80°F, then stratify for 2 to 4 months at 33 to 41°F.
6 = Scarify (thin by abrasion) seedcoat or soak seed in concentrated sulfuric acid for 20 to 120 minutes.
7 = Stratify for 1 to 2 months at 33 to 41°F.
8 = Stratify for 6 months at 41°F.
9 = Stratify for 1 month at 34°F.

APPENDIX C

ESTIMATING THE ROUGHNESS COEFFICIENT (n)

The roughness coefficient (n)[1] is used in Manning's formula to account for anything that retards water flow in the channel. Retardation can be caused by grass, weeds, other vegetation, channel crookedness, variations in channel shape, and obstructions such as roots, stumps, stones, or any other debris. It is important that the value selected for *n* be valid. The value of 0.022 often used in this manual is valid for an average diversion ditch channel cut in a minesoil, but exceptions to this can be evaluated by using the following method.

The first step is to select a basic value for *n* from Table C-1. Select the value that best fits the material in which the channel is being cut.

Next (step 2), modify the basic value by adding the most appropriate value from Table C-2. Consider the material to be used, and select the degree of surface smoothness possible under the material limitations.

Step 3 is evaluating the variations in shape and size of the channel (Table C-3). Look for abrupt changes in shape or size. These changes cause more turbulence than gradual changes.

Table C-1. Basic *n* Value

Character of channel	Basic *n* value
Channels in earth	0.02
Channels cut into rock	0.025
Channels in fine gravel	0.024
Channels in coarse gravel	0.028

[1]U.S. Department of Agriculture, Soil Conservation Service. 1974. Erosion and Sediment Control in Developing Areas--Planning Guidelines and Design Aids. Columbia, S.C.

Table C-2. Modifying Value for Surface Irregularity

Degree of irregularity	Surfaces	Modifying value
Smooth	Best obtainable for materials involved	0.000
Minor	Good graded channels; slightly eroded or scoured side slopes of canals or drainage channels	0.005
Moderate	Fair to poorly graded channels; moderately sloughed or eroded side slopes of canals or drainage channels	0.010
Severe	Very poorly formed channels; badly sloughed banks of natural channels; badly eroded or sloughed sides of canals or drainage channels; unshaped, jagged and irregular surfaces of channels excavated in rock	0.020

Step 4 is the evaluation of obstructions found in the channel, such as roots, stumps, logs, debris tangles, and large rocks. Remember that sharp-edged materials cause more turbulence than smooth obstructions. Use Table C-4 for this step and be sure not to evaluate anything in this step that was evaluated in another step.

Step 5 is the evaluation of the crookedness or meandering of the channel. To obtain this value measure the part of the channel that is crooked (l_m), and then divide this length by the length that is straight (l_s) or smoothly curving. Use this value in Table C-5 for the crookedness or meander value.

Step 6, the final step, is an evaluation of the effect of vegetation on the n value. Use Table C-6 to obtain this value. This value will

Table C-3. Modifying Value for Variations of Cross Sections

Character of variations in size and shape of cross sections	Modifying value
Changes in size or shape occur gradually	0.000
Large and small sections alternate occasionally or shape changes cause occasional shifting of main flow from side to side	0.005
Large and small sections alternate frequently or shape changes cause frequent shifting of main flow from side to side	0.010 to 0.015

Table C–4. Modifying Value for Obstructions

Relative effect of obstructions	Modifying value
Negligible	0.000
Minor	0.010 to 0.015
Appreciable	0.020 to 0.030
Severe	0.040 to 0.060

probably be 0 on most minesoils because the diversion ditches are built and used before vegetation can be established in the ditches.

> **Example:** Assume that a diversion ditch has been constructed in a sandy clay loam with many roots and limbs left in the replaced topsoil. These roots and limbs interfered with construction and caused some irregularity in the channel sides and bottom. Changes in the channel shape and size occur gradually or not at all. There is no vegetation in the channel. The channel is 300 feet long, of which 250 feet are constructed in a smooth and steady curve and 50 feet are crooked channel.

C–1	Basic n (earth)	0.020
C–2	Irregularity (moderate)	0.010
C–3	Variations (almost none)	0.000
C–4	Obstructions (appreciable)	0.020
C–5	Meander ($\frac{l_m}{l_s} = \frac{50\ \text{ft}}{250\ \text{ft}} = 0.2$)	0.000
C–6	Vegetation (none)	0.000
	$n =$	0.050

Table C–5. Modifying Value for Meander

Ratio l_m/l_s	Degree of meandering	Modifying value
0.0 to 1.2	Minor	0.000
1.2 to 1.5	Appreciable	0.15
1.5 and greater	Severe	0.30

Table C-6. Modifying Value for Vegetation

Vegetation and flow conditions	Degree of effect on *n* value	Range in modifying value
Dense growths of flexible turf grasses or weeds (eg, Bermuda and blue grasses) where average depth of flow is two to three times height of vegetation Supple seedling tree switches (eg, willow, cottonwood, salt cedar) where average depth of flow is three to four times height of vegetation	Low	0.005 to 0.010
Turf grasses where average depth of flow is one to two times height of vegetation Stemmy grasses, weeds, or tree seedlings with moderate cover where average depth of flow is two to three times height of vegetation Moderately dense brushy growths, similar to 1- to 2-year-old willows, dormant season, along side slopes of channel with no significant vegetation along the channel bottom, where the hydraulic radius is greater than 2 feet	Medium	0.010 to 0.025
Turf grasses where the average depth of flow is about equal to the height of vegetation Dormant season, 8- to 10-year-old willow or cottonwood trees, intergrown with some weeds and brush, no vegetation in foliage, where the hydraulic radius is greater than 2 feet Growing season, about 1-year-old bushy willows intergrown with some weeds in full foliage along side slopes, no significant vegetation along channel bottom, where hydraulic radius is greater than 2 feet	High	0.025 to 0.050

continued

Table C–6. *(Continued)*

Vegetation and flow conditions	Degree of effect on *n* value	Range in modifying value
Turf grasses where the average depth of flow is less than one-half height of vegetation Growing season, about 1-year-old bushy willows intergrown with weeds in full foliage along side slopes; dense growth of cattails along channel bottom; any value of hydraulic radius up to 10 or 15 feet Growing season; trees intergrown with weeds and brush, all in full foliage; any value of hydraulic radius up to 10 or 15 feet	Very high	0.050 to 0.100

APPENDIX D

NOMOGRAPH FOR SOLVING UNIVERSAL SOIL LOSS EQUATION

The nomograph in Figure D–1 can be used to solve for any value (*A, R, K, LS*) in the equation $A = RKLS$ if the other three values are known.

Example: Assume that an area has a rainfall factor *(R)* of 165, a soil erodibility factor *(K)* of 0.33, and a topographic factor *(LS)* of 2.0. We would like to know the erosion potential *(A)* for this area in tons of soil per acre per year. Draw a straight line from 165 on the *R* line to 0.33 on the *K* line. Now draw a line from the point where this first line intersects the turning axis line to 2.0 on the *LS* line. The point at which this second line intersects the *A* line is the erosion potential. The erosion potential is approximately 110 tons per acre per year if the values for *C* and *P* are one. Any other values of *C* and *P* can be multiplied against 110 tons per acre per year.

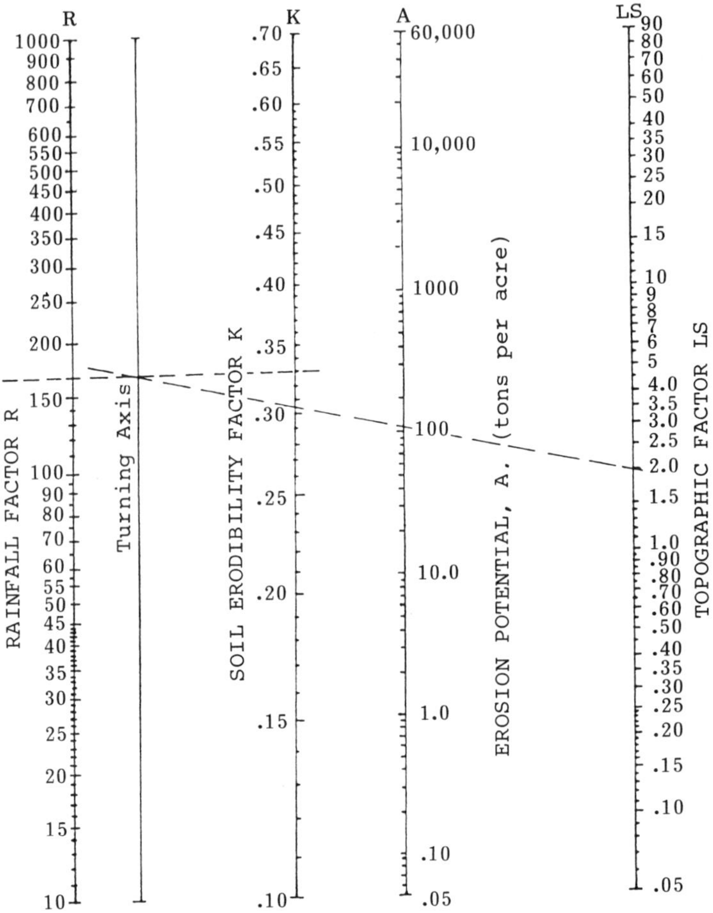

Figure D-1. Nomograph for solving the universal soil loss equation with $A = RKLS$. C and P have values of 1.0.

APPENDIX E

MANNING'S FORMULA

Manning's formula is an empirical equation expressing the mean velocity flow as a function of channel roughness, hydraulic radius, and channel slope.

The four nomographs on the following pages (Figures E-1 through E-4) can be used to estimate any value of the equation if the other three equation values are known. Each nomograph is designed to be used for a specific range of values of hydraulic radius (r), roughness coefficient (n), channel slope (s), and mean velocity (v).

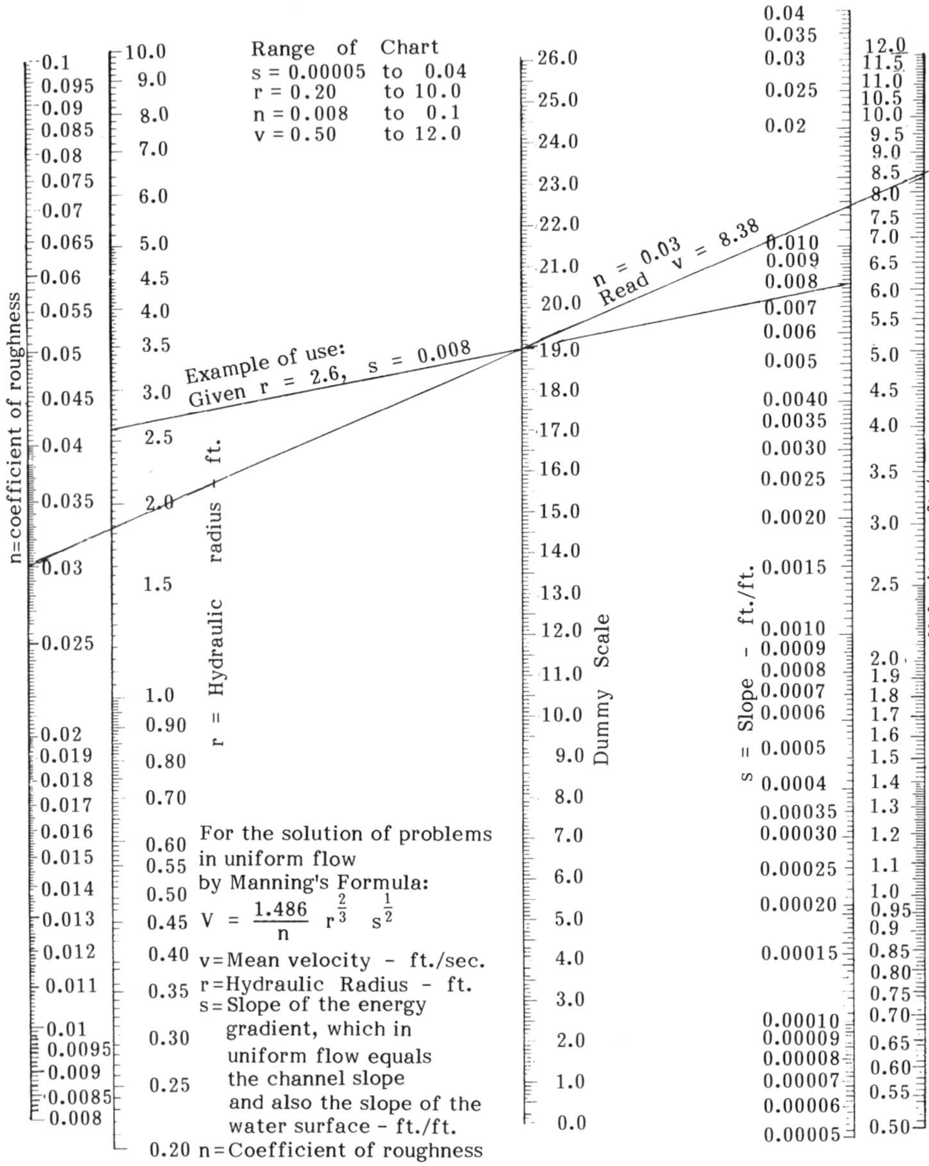

Figure E–1. Nomograph for solving Manning's formula. *Source:* U.S. Department of Agriculture, Soil Conservation Service. 1979. *S.C.C. Engineering Field Manual for Conservation Practices.* Washington, D.C.: U.S. Government Printing Office.

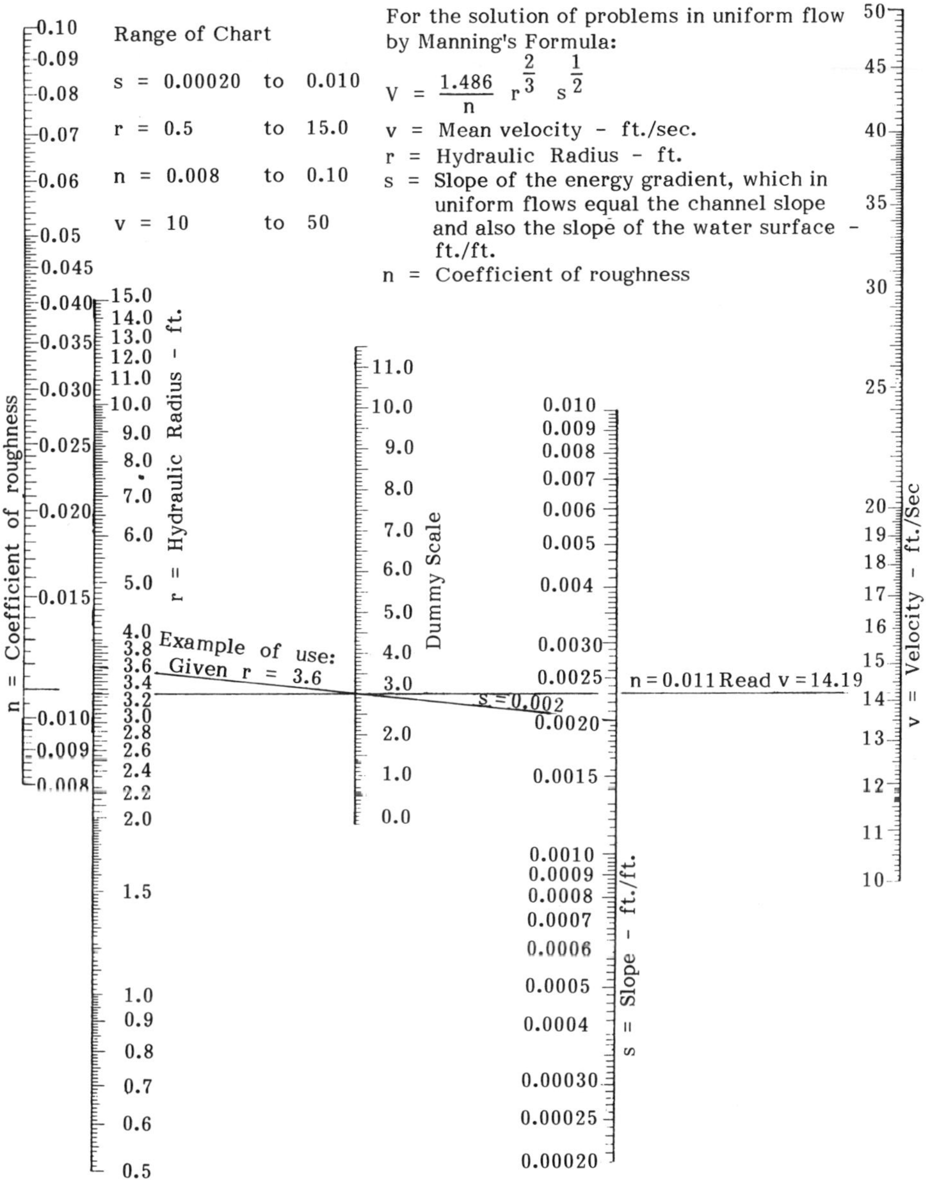

Figure E–2. Nomograph for solving Manning's formula. *Source:* U.S. Department of Agriculture, Soil Conservation Service. 1979. *S.C.S. Engineering Field Manual for Conservation Practices.* Washington, D.C.: U.S. Government Printing Office.

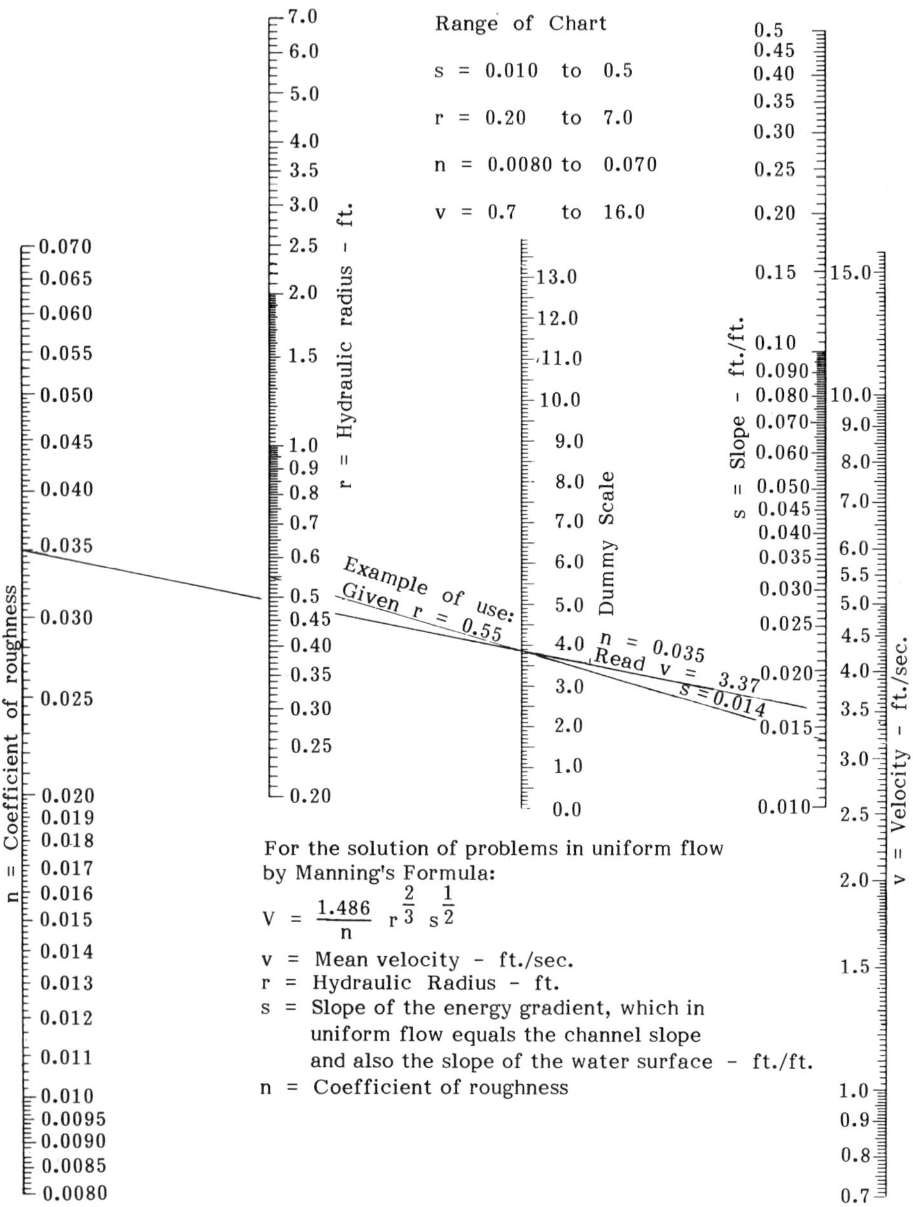

Figure E–3. Nomograph for solving Manning's formula. *Source:* U.S. Department of Agriculture, Soil Conservation Service. 1979. *S.C.S. Engineering Field Manual for Conservation Practices.* Washington, D.C.: U.S. Government Printing Office.

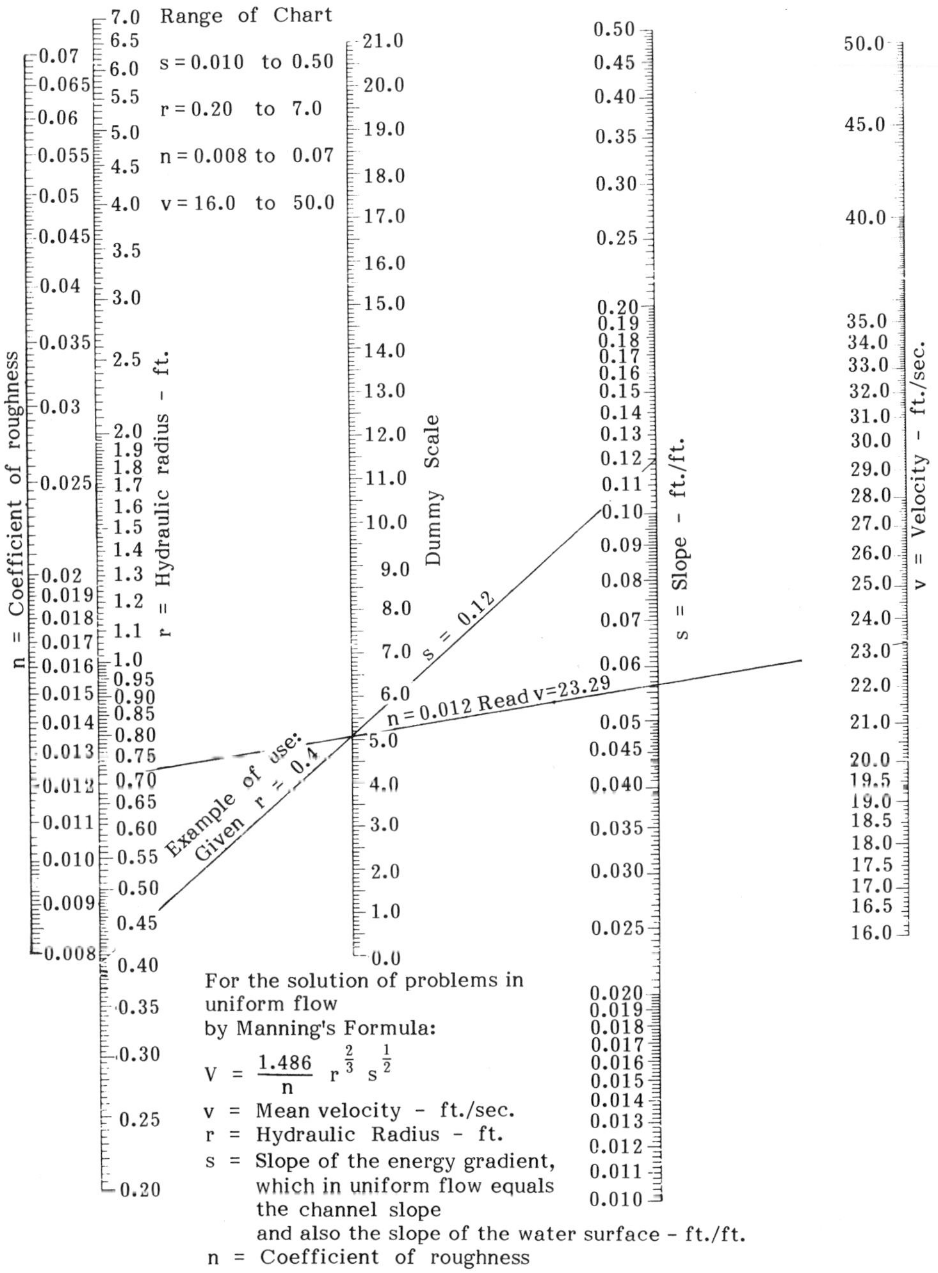

Figure E–4. Nomograph for solving Manning's formula. *Source:* U.S. Department of Agriculture, Soil Conservation Service. 1979. *S.C.S. Engineering Field Manual for Conservation Practices.* Washington, D.C.: U.S. Government Printing office.

APPENDIX F

TRAPEZOIDAL CHANNEL OUTLET DESIGN

To use this method of outlet design,[1] the expected peak flow and the channel slope are needed. Two graphs have been developed based on these two values. One of the graphs (Figure F-1) shows the riprap size needed for the narrowest channel that would be practical. The other graph (Figure F-2) shows the riprap size needed for the widest channel that would be practical. There are many acceptable channel designs between these two extremes. Also, the riprap sizes shown for the two extreme designs are the minimum sizes that can be used successfully, and larger sizes will perform as well or better.

The riprap sizes shown in Figure F-1 and F-2 are designated as d_{50}. This d_{50} value means that 50 percent by weight of the riprap should be smaller in diameter. The riprap size and the slope can now be used with Figure F-3 to obtain the water velocity in the channel at peak flow.

The slope and riprap size can be used again with Figure F-4 to obtain the hydraulic radius of the channel cross section. The hydraulic radius and the cross-section area of the channel are needed to obtain the depth and bottom width of the channel. The area is obtained from Figure F-5 based on velocity and peak flow.

The next value needed to design a channel is the slope of the channel sides. To obtain slope, the shape of the rock to be used must be considered. Figure F-6 shows the angle of repose (θ) that can be expected for rock of three different general shapes. This angle is the angle between the horizontal and the maximum slope at which the rock will remain stable. To use Figure F-6, a decision must be made as to

[1]This method is taken from Anderson, A.G., A.S. Paintel, and J.T. Davenport. 1970. Tentative Design Procedure for Riprap-Lined Channels, National Cooperative Highway Research Program Report No. 108. National Academy of Sciences. Washington, D.C.

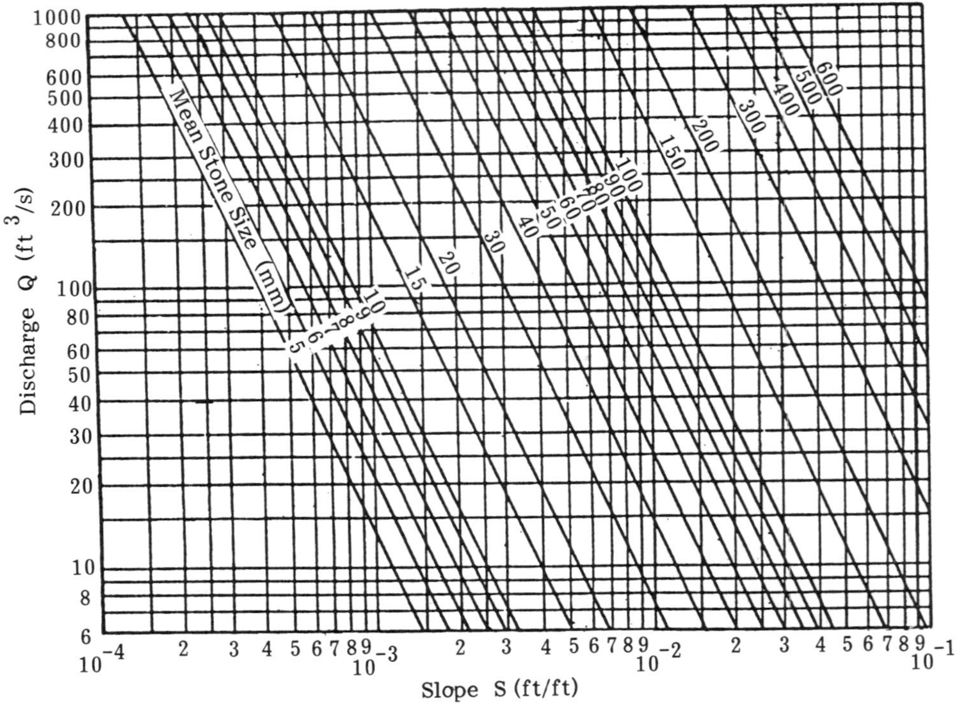

Figure F–1. Mean minimum size of stone riprap that will be stable in the narrowest trapezoidal channels for combinations of discharge (Q) and slope (S). *Source:* Anderson, A.G.. A.S. Paintel, and J.T. Davenport. 1970. *Tentative Design Procedure for Riprap-Lined Channels,* National Cooperative Highway Research Program Report No. 108. National Academy of Sciences. Washington, D.C.

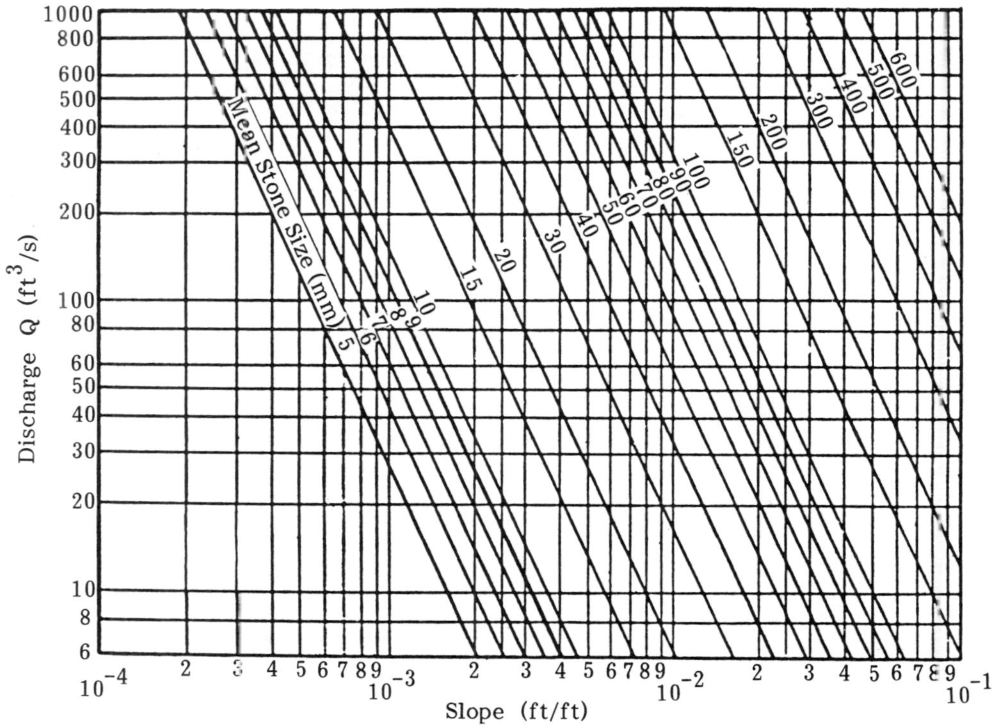

Figure F–2. Mean minimum size of stone riprap that will be stable in the widest trapezoidal channels for combinations of discharge (Q) and slope (S). *Source:* Anderson, A.G., A.S. paintel, and J.T. Davenport. 1970. *Tentative Design Procedure for Riprap-Lined Channels,* National Cooperative Highway Research Program Report No. 108. National Academy of Sciences. Washington, D.C.

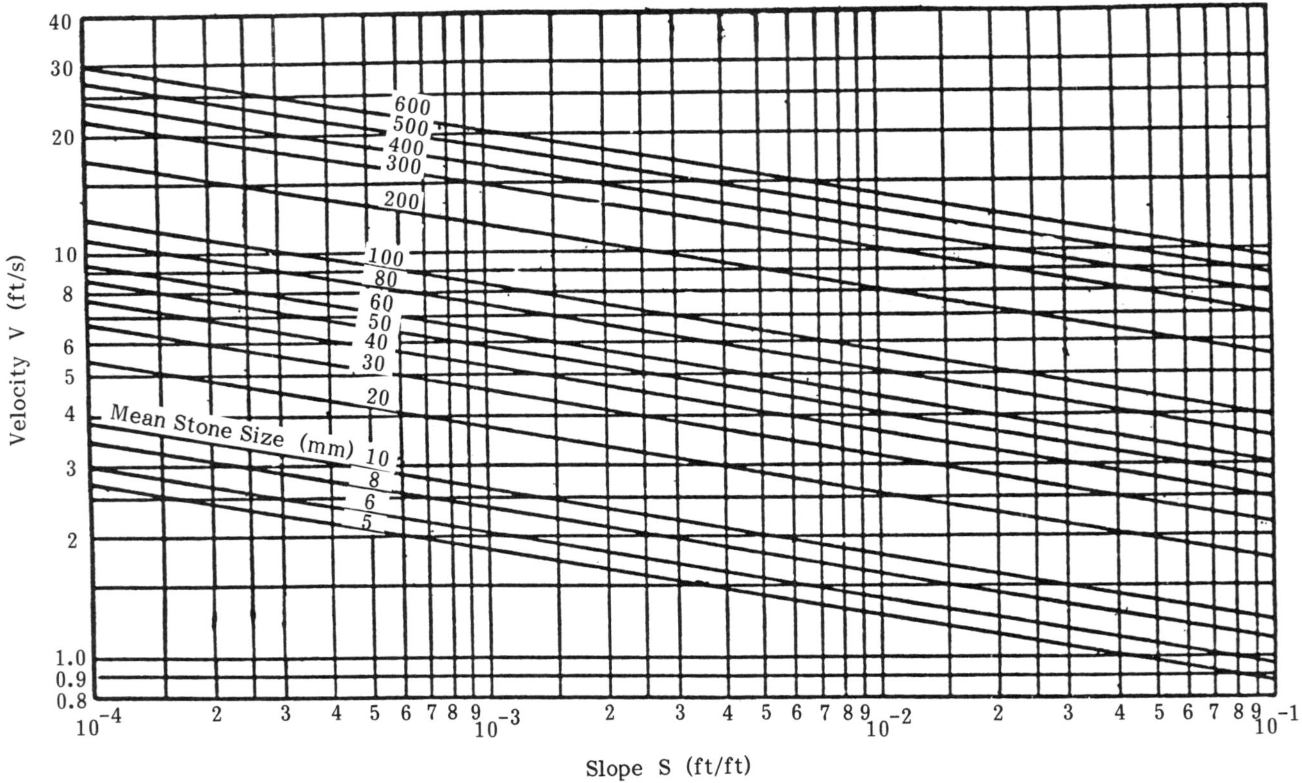

Figure F-3. Maximum mean velocity (*V*) for stable riprap in trapezoidal channels for various mean stone sizes and slopes (*S*). *Source:* Anderson, A.G., A.S. Paintel, and J.T. Davenport. 1970. *Tentative Design Procedure for Riprap-Lined Channels,* National Cooperative Highway Research Program Report No. 108. National Academy of Sciences. Washington, D.C.

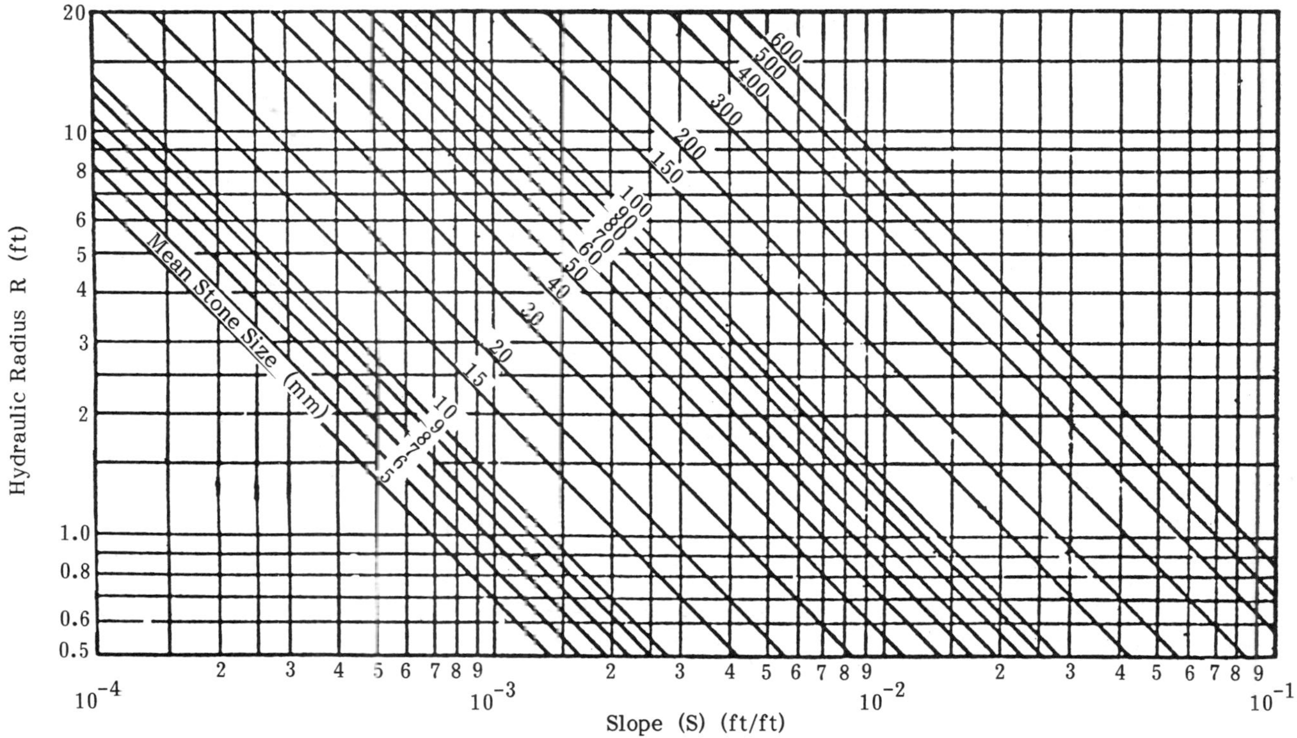

Figure F-4. Hydraulic radius (*R*) for trapezoidal channels in terms of stone size and slope (*S*). *Source:* Anderson, A.G., A.S. Paintel, and J.T. Davenport. 1970. *Tentative Design Procedure for Riprap-Lined Channels,* National Cooperative Highway Research Program Report No. 108. National Academy of Sciences. Washington, D.C.

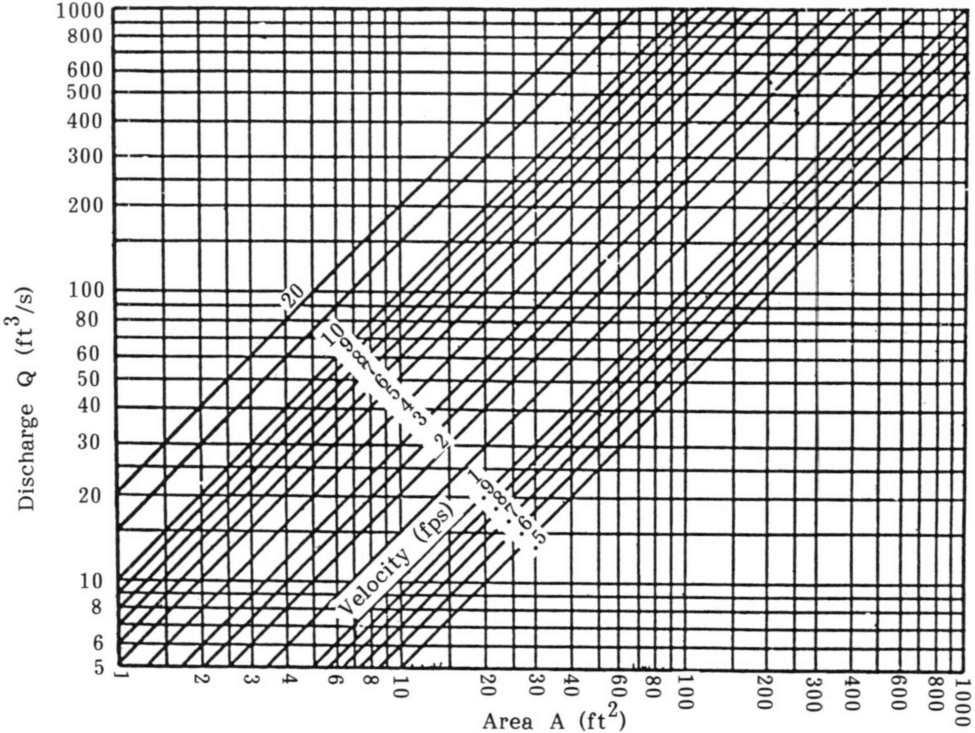

Figure F–5. Cross-section area (A) of a trapezoidal channel in terms of discharge (Q) and maximum mean velocity (V). *Source:* Anderson, A.G., A.S. Paintel, and J.T. Davenport. 1970. *Tentative Design Procedure for Riprap-Lined Channels,* National Cooperative Highway Research Program Report No. 108. National Academy of Sciences. Washington, D.C.

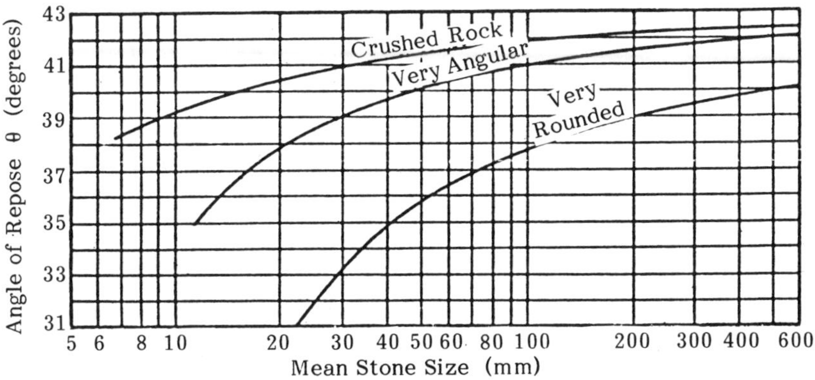

Figure F–6. Angle of repose (θ) of riprap in terms of size and shape of stone. *Source:* Anderson, A.G., A.S. Paintel, and J.T. Davenport. 1970. *Tentative Design Procedure for Riprap-Lined Channels,* National Cooperative Highway Research Program Report No. 108. National Academy of Sciences. Washington, D.C.

whether the rock is round in shape, was obtained by crushing, or has a shape between these two extremes. The angle of repose is used with Figure F–7 to determine whether the side slope can be 4:1, 3:1, 2.5:1, or a lesser slope. If the angle of repose is over 50°, side slopes of 2:1 and 1.5:1 can be used.

The channel cross-section area and hydraulic radius are now known and can be used to determine the bottom width and depth of the channel. These values are obtained from Figures F–8, F–9, F–10, F–11, and F–12, depending on the side slope to be used. The depth of channel should be increased by a factor of 0.25 to compensate for construction irregularities. The depth of channel, width of bottom, side slope, and size of riprap are now known for building a successful outlet.

The average rock size (d_{50}) for the riprap has been determined but not the maximum or minimum size. The maximum size should be 1.5 times the average size (d_{50}). The minimum size can be anything down to sand size, but should be no greater than 1 inch in diameter. The rock sizes should be well graded from the largest size down to the smallest size. Under these circumstances the voids between the larger rocks will be filled with the smaller rocks and flowing water will not erode the soil beneath the riprap.

The minimum thickness of the riprap layer should be at least 1.5 times the maximum rock size. However, the layer should not be less than 6 inches thick. Another method of selecting thickness is to multiply average rock size (d_{50}) by three. This will result in a thicker riprap layer than the first method.

Riprap should be spread uniformly so that the layer is homogenous in regard to rock sizes. The effectiveness of the riprap can be increased by compacting the rock with a roller or other device. Compacted rock

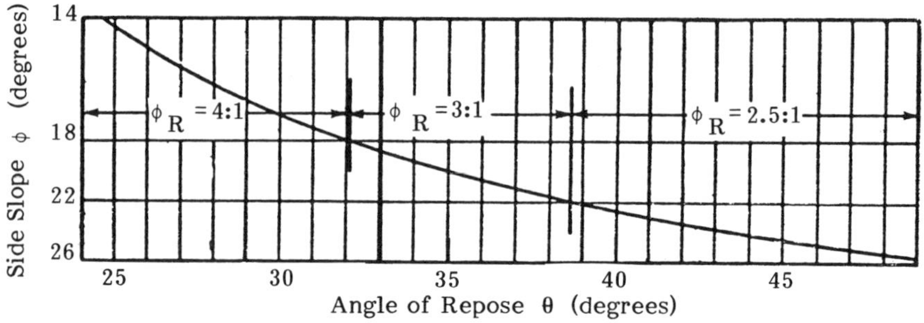

Figure F-7. Side slopes (ϕ) to be used for various angles of repose (θ). *Source:* Anderson, A.G., A.S. paintel, and J.T. Davenport. 1970. *Tentative Design Procedure for Riprap-Lined Channels,* National Co-operative Highway Research Program Report No. 108. National Academy of Sciences. Washington, D.C.

is also more resistant to displacement if vehicles must pass over the riprap area.

Example: An outlet is needed to carry the water from a series of diversion ditches to a sedimentation pond. The outlet will have to be lined with riprap because there is insufficient time to establish vegetation on the soil lining the outlet channel. The peak flow expected in a 10-year period for a 24-hour duration storm is 400 cubic feet per second (cfs) and the slope of the outlet channel is 0.05 foot per foot (ft/ft). The rock available for riprap is a crushed material.

Figure F-1: minimum average size = 600 mm (23.6 in.)

Figure F-3: velocity = 12 ft/s

Figure F-4: hydraulic radius = 1.7 ft

Figure F-6: channel cross-section area = 38 ft^2

Figure F-6: angle of repose (θ) = 42.3°

Figure F-7: side slope (Φ) = 2.5:1 (23.5°)

Figure F-10: bottom width = 10 ft
water depth = 2.5 ft

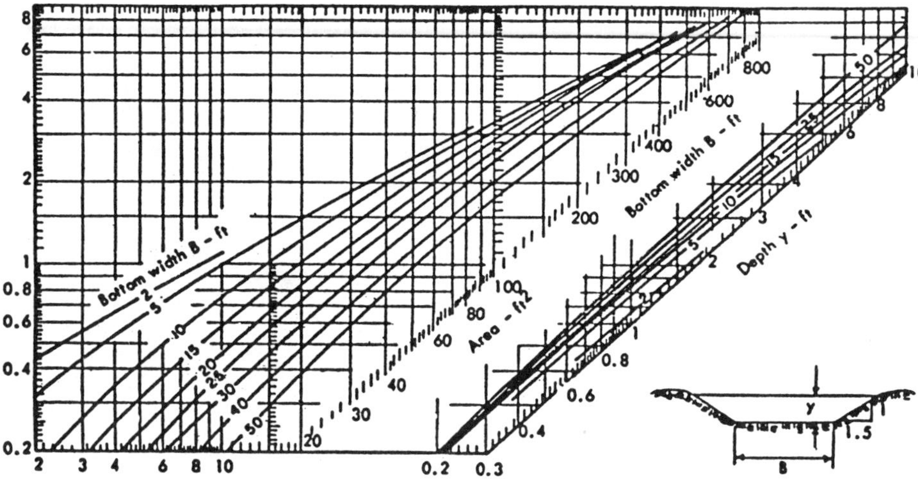

Figure F–8. Bottom widths (B) and depths of flow (Y) for trapezoidal channels with 1.5:1 side slopes. *Source:* Anderson, A.G., A.S. paintel, and J.T. Davenport. 1970. *Tentative Design Procedure for Riprap-Lined Channels,* National Cooperative Highway Research Program Report No. 108. National Academy of Sciences. Washington, D.C.

This channel would have the narrowest practical channel and the largest rock. The other channel extreme would be more shallow with smaller rock:

Figure F–2: minimum average size = 450 mm (17.7 in.)

Figure F–3: velocity = 9.2 ft/s

Figure F–4: hydraulic radius = 1.3 ft

Figure F–5: channel cross-section area = 46 ft^2

Figure F–6: angle of repose (θ) = 42.3°

Figure F–7: side slope (Φ) = 2.5:1 (23.5°)

Figure F–10: bottom width = 25 ft
water depth = 1.6 ft

Assume that both of these channels are unsuitable for some reason, such as the rock size. Remember that you can use any d_{50} between 600 mm and 450 mm. For example, if there was a good source of rock that would fit the d_{50} size of 500 mm (19.7 in.), you could recalculate the channel dimensions to fit this rock size:

Figure F–3: velocity = 9.7 ft/s

Figure F–4: hydraulic radius = 1.2 ft

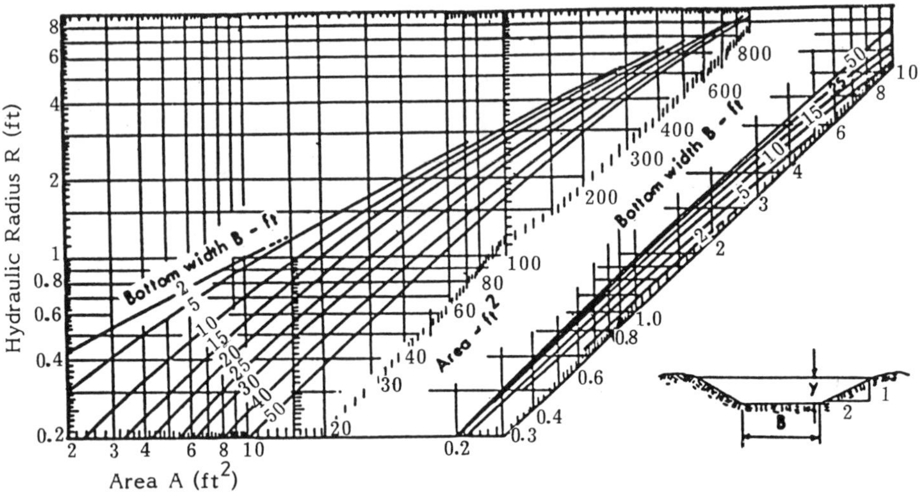

Figure F-9. Bottom widths (B) and depths of flow (Y) for trapezoidal channels with 2.0:1 side slopes. *Source:* Anderson, A.G., A.S. Paintel, and J.T. Davenport. 1970. *Tentative Design Procedure for Riprap-Lined Channels,* National Cooperative Highway Research Program Report No. 108. National Academy of Sciences. Washington, D.C.

Figure F-5:	channel cross-section area = 42 ft^2
Figure F-6:	angle of repose (θ) = 42.3°
Figure F-7:	side slope (Φ) = 2.5:1 (23.5°)
Figure F-10:	bottom width = 25 ft water depth = 1.4 ft

If this channel is chosen, the riprap thickness will be:

1.5 x d_{50} = maximum rock diameter

1.5 x 19.7 in. = 29.5 in.

1.5 x maximum rock diameter = riprap thickness

1.5 x 29.5 in. = 44.25 in.

The calculated water depth should be increased by 0.25 to obtain channel depth. However, the channel depth should be at least 1 foot greater than the calculated water depth.

$$0.25 \times 1.4 \text{ ft} = 0.35 \text{ ft}$$

As this (0.35 ft) is less than 1.0 ft, the channel depth should be increased by 1.0 ft:

$$1.4 \text{ ft} + 1.0 \text{ ft} = 2.4 \text{ ft}$$

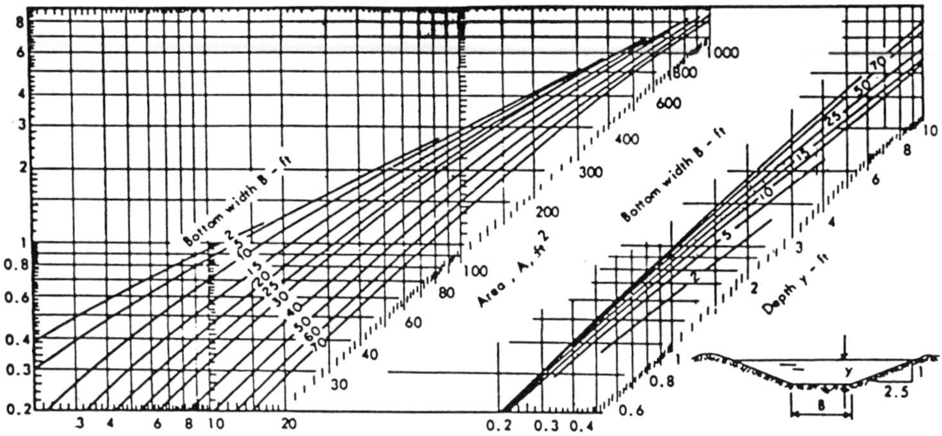

Figure F-10. Bottom widths (*B*) and depths of flow (*Y*) for trapezoidal channels with 2.5:1 side slopes. *Source:* Anderson, A.G., A.S. Paintel, and J.T. Davenport. 1970. *Tentative Design Procedure for Riprap-Lined Channels,* National Cooperative Highway Research Program Report No. 108. National Academy of Sciences. Washington, D.C.

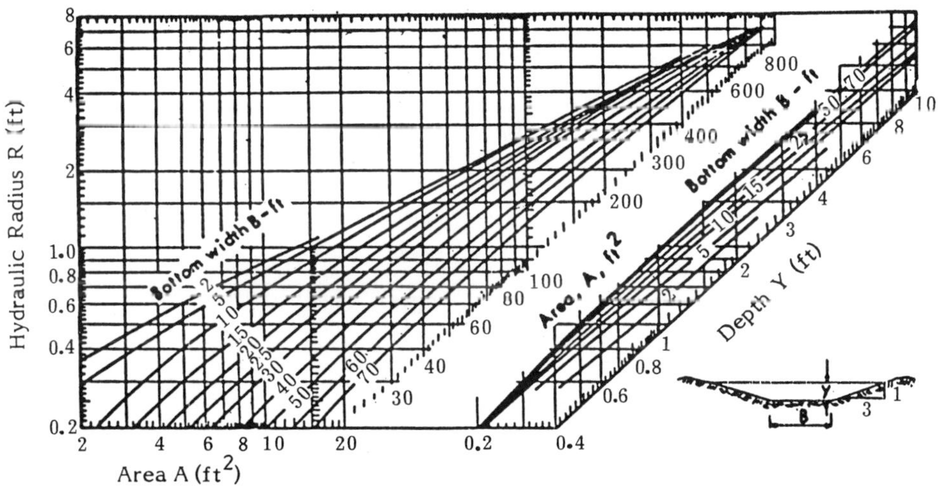

Figure F-11. Bottom widths (*B*) and depths of flow (*Y*) for trapezoidal channels with 3.0:1 side slopes. *Source:* Anderson, A.G., A.S. Paintel, and J.T. Davenport. 1970. *Tentative Design Procedure for Riprap-Lined Channels,* National Cooperative Highway Research Program Report No. 108. National Academy of Sciences. Washington, D.C.

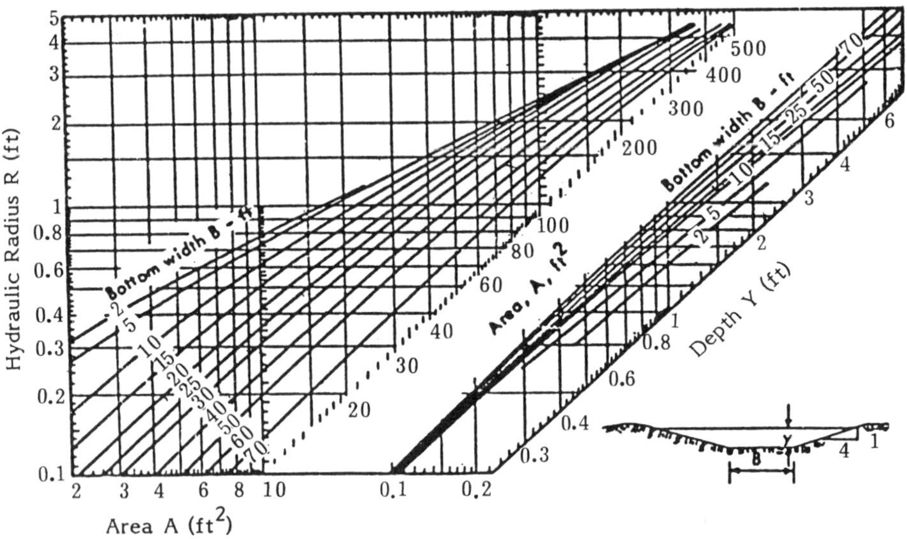

Figure F-12. Bottom widths (B) and depths of flow (Y) for trapezoidal channels with 4.0:1 side slopes. *Source:* Anderson, A.G., A.S. Paintel, and J.T. Davenport. 1970. *Tentative Design Procedure for Riprap-Lined Channels,* National Cooperative Highway Research Program Report No. 108. National Academy of Sciences. Washington, D.C.

INDEX